Next-Time Questions
to accompany

CONCEPTUAL
Physics
SEVENTH EDITION

City College of San Francisco

HarperCollinsCollegePublishers

USE THESE MASTERS TO MAKE COPIES FOR POSTING, OR TO MAKE TRANSPARENCIES FOR OVERHEAD PROJECTION. END YOUR CLASS WITH A QUESTION --- THEN BEGIN THE NEXT WITH AN ANSWER.

THERE IS A TOTAL OF 128

ENJOY!

Next-Time Questions to accompany *CONCEPTUAL PHYSICS*, Seventh Edition

Copyright © 1993 by Paul G. Hewitt

All rights reserved. Printed in the United States of America. For information address HarperCollins College Publishers, 10 E. 53rd St., New York, NY 10022.

ISBN 0-673-54147-9

94 95 5 4 3 2

CONCEPTUAL Physics

"WHICH OF THE STATEMENTS BELOW IS A SCIENTIFIC CLAIM?"

1. HUMAN BEINGS WILL NEVER SET FOOT ON THE MOON.

2. SOME OF THE LAWS THAT GOVERN NATURE CANNOT BE DETECTED BY SCIENTISTS.

3. IT IS QUITE POSSIBLE THAT IN SOME OTHER GALAXY THE LAWS OF PHYSICS ARE FUNDAMENTALLY DIFFERENT THAN THE LAWS WE ARE ACQUAINTED WITH IN THIS GALAXY.

CONCEPTUAL Physics

WHICH OF THE STATEMENTS BELOW IS A SCIENTIFIC CLAIM?

1. Human beings will never set foot on the moon.
2. Some of the laws that govern nature cannot be detected by scientists.
3. It is quite possible that in some other galaxy the laws of physics are fundamentally different than the laws we are acquainted with in this galaxy.

ANSWER:

Only statement 1 is scientific, because there is a test for its wrongness. The claim not only is capable of being proved wrong, it in fact HAS been proved wrong. So even though the statement is incorrect, it is nonetheless scientific.

Statement 2 has no test for its possible wrongness, and is therefore unscientific.

Likewise with statement 3, which is speculation. Finding one galaxy with different laws would prove the statement true, but what is the test for proving it wrong? If we searched the universe and found no galaxies with different laws, this would not be proof that a galaxy "just around the corner" doesn't operate under different laws. A claim that is capable of being proved right, but not capable of being proved wrong is not a scientific claim.

One who makes a scientific claim deliberately places oneself at risk of admitting the claim is wrong --- the scientist says, "IF YOU CONDUCT A TEST AND IT TURNS OUT TO BE NEGATIVE, THEN MY CLAIM IS WRONG."

To be a scientist, you must gracefully accept the outcome of a test, whether positive or negative.

THAT'S THE SPIRIT OF INQUIRY!

CONCEPTUAL Physics

A motorist wishes to travel 40 kilometers at an average speed of 40 km/h. During the first 20 kilometers, an average speed of 40 km/h is maintained. During the next 10 kilometers, however, the motorist goofs off and averages only 20 km/h. To drive the last 10 kilometers and average 40 km/h the motorist must drive

a) 60 km/h

b) 80 km/h

c) 90 km/h

d) FASTER THAN THE SPEED OF LIGHT

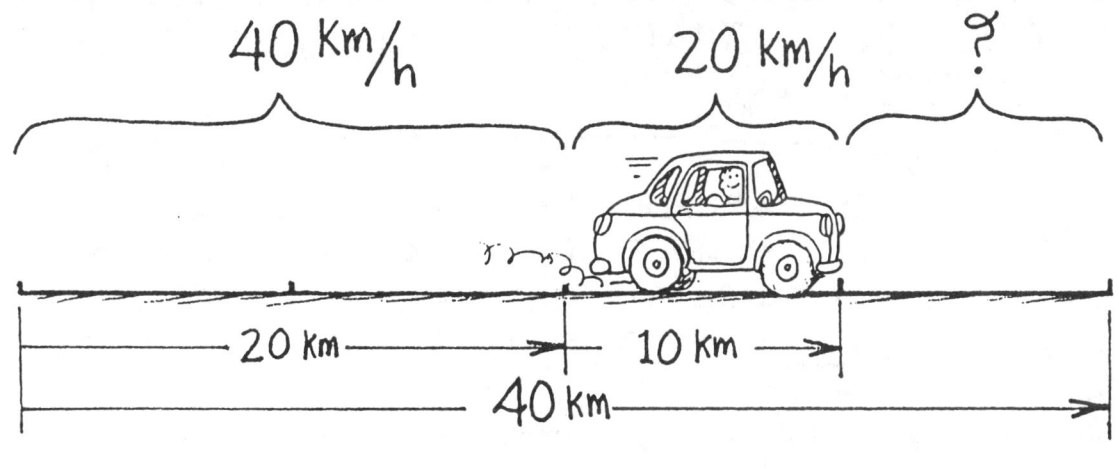

CONCEPTUAL Physics

A motorist wishes to travel 40 kilometers at an average speed of 40 km/h. During the first 20 kilometers, an average speed of 40 km/h is maintained. During the next 10 kilometers, however, the motorist goofs off and averages only 20 km/h. To drive the last 10 kilometers and average 40 km/h the motorist must drive

a) 60 km/h
b) 80 km/h
c) 90 km/h
d) faster than the speed of light

THE ANSWER IS d:

You would have to travel at an infinite speed and finish the last 10 kilometers in zero time to attain an average speed of 40 km/h! Why? Because you have one hour to make the trip, and your one hour is up at the 30-kilometer point. You spent ½ hour to the half-way point, 20 kilometers, and another ½ hour when you averaged 20 km/h over that 10-kilometer stretch. So you'd have to cover the entire 40 kilometers in one hour -- that means, the last 10 kilometers in no time at all.

Be careful in averaging speeds like you average distances. Speed involves distance and TIME. Be sure to consider time in problems that involve speed!

CHAP. 2

CONCEPTUAL Physics

WHEN THE 10 km/h BIKES ARE 20 km APART, A BEE BEGINS FLYING FROM ONE WHEEL TO THE OTHER AT A STEADY SPEED OF 30 km/h. WHEN IT GETS TO THE WHEEL, IT ABRUPTLY TURNS AROUND AND FLIES BACK TO TOUCH THE FIRST WHEEL, THEN TURNS AROUND AND KEEPS REPEATING THE BACK-AND-FORTH TRIP UNTIL THE BIKES MEET, AND ≈SQUISH!≈

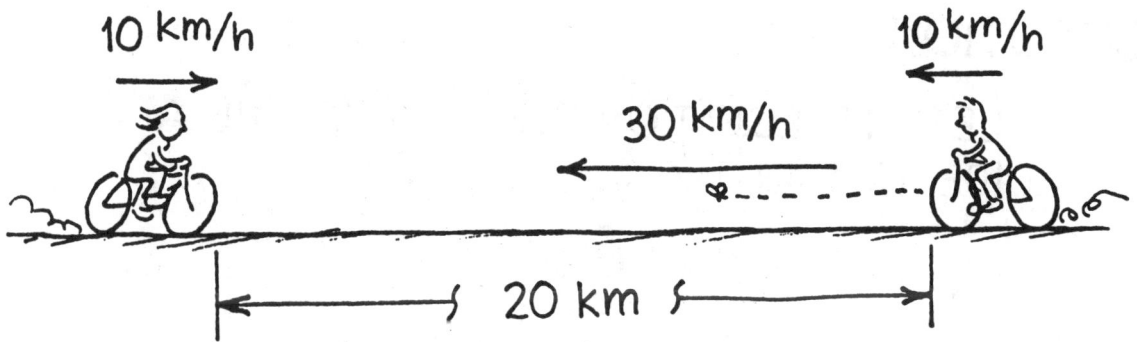

QUESTION

HOW MANY KILOMETERS DID THE BEE TRAVEL IN ITS TOTAL BACK-AND-FORTH TRIPS?

CHAP. 2

CONCEPTUAL Physics

WHEN THE 10 km/h BIKES ARE 20 km APART, A BEE BEGINS FLYING FROM ONE WHEEL TO THE OTHER AT A STEADY SPEED OF 30 km/h. WHEN IT GETS TO THE WHEEL, IT ABRUPTLY TURNS AROUND AND FLIES BACK TO TOUCH THE FIRST WHEEL, THEN TURNS AROUND AND KEEPS REPEATING THE BACK-AND-FORTH TRIP UNTIL THE BIKES MEET, AND ⸘SQUISH‼⸘

QUESTION

HOW MANY KILOMETERS DID THE BEE TRAVEL IN ITS TOTAL BACK-AND-FORTH TRIPS?

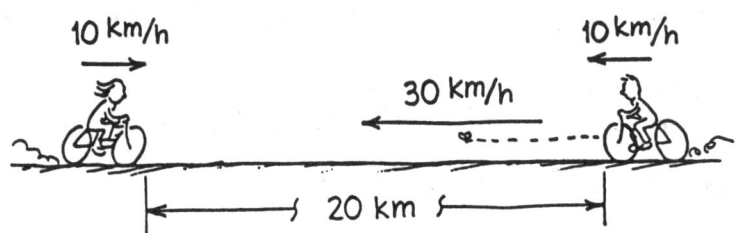

SOLUTION:

LET THE EQUATION FOR DISTANCE BE A GUIDE TO THINKING:

$$d = \bar{v}\, t$$

WE KNOW $\bar{v} = 30$ km/h, AND WE MUST FIND THE TIME t. WE CONSIDER THE SAME TIME FOR THE BIKES AND SEE IT TAKES 1 HOUR FOR THEM TO MEET, SINCE EACH TRAVELS 10 km AT A SPEED OF 10 km/h. SO,

$$d = \bar{v}\, t = 30 \text{ km/h} \times 1 \text{ h} = 30 \text{ km}$$

THE BEE TRAVELED A TOTAL OF 30 km.

CHAP. 2

CONCEPTUAL Physics

THE BOY ON THE TOWER THROWS A BALL 20 METERS DOWNRANGE AS SHOWN.

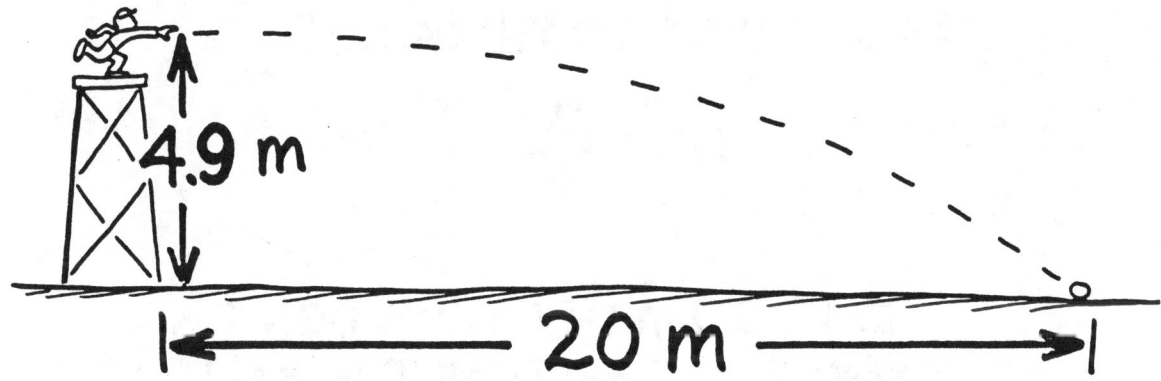

WHAT IS HIS PITCHING SPEED?

CHAP. 3

CONCEPTUAL Physics

THE BOY ON THE TOWER THROWS A BALL 20 METERS DOWNRANGE AS SHOWN.

WHAT IS HIS PITCHING SPEED?

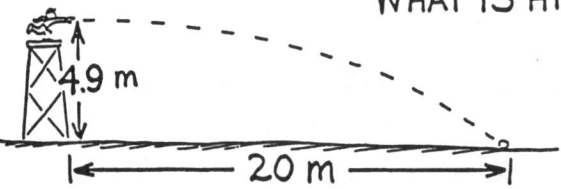

SOLUTION

USE THE EQUATION FOR SPEED AS A "GUIDE TO THINKING".

$$v = \frac{d}{t}$$

d IS 20 m; BUT WE DON'T KNOW t... THE TIME THE BALL TAKES TO GO 20 m. BUT WHILE THE BALL MOVES HORIZONTALLY 20 m, IT FALLS A VERTICAL DISTANCE OF 4.9 m, WHICH TAKES 1 SECOND... SO $t = 1$ s.

$$v = \frac{d}{t} = \frac{20 \text{ m}}{1 \text{ s}} = 20 \text{ m/s}$$

CONCEPTUAL **Physics**

A ZOOKEEPER DEVISES A RUBBER-BAND GUN TO SHOOT FOOD TO A MONKEY WHO IS TOO SHY TO COME DOWN FROM THE TREES.

IF THE MONKEY DOES NOT MOVE, SHOULD THE KEEPER AIM ABOVE, AT, OR BELOW THE MONKEY?

IF THE MONKEY LETS GO OF THE BRANCH AT THE INSTANT THE KEEPER SHOOTS THE FOOD, SHOULD THE KEEPER AIM ABOVE, AT, OR BELOW THE MONKEY TO GET FOOD TO THE MONKEY IN MID-AIR?

BANANA

CHAP. 3

CONCEPTUAL Physics

A ZOOKEEPER DEVISES A RUBBER-BAND GUN TO SHOOT FOOD TO A MONKEY WHO IS TOO SHY TO COME DOWN FROM THE TREES.

IF THE MONKEY DOES NOT MOVE, SHOULD THE KEEPER AIM ABOVE, AT, OR BELOW THE MONKEY?

IF THE MONKEY LETS GO OF THE BRANCH AT THE INSTANT THE KEEPER SHOOTS THE FOOD, SHOULD THE KEEPER AIM ABOVE, AT, OR BELOW THE MONKEY TO GET FOOD TO THE MONKEY IN MID-AIR?

BANANA

LINE OF SIGHT (DASHED LINE)

FOR RELATIVELY HIGH-SPEED SHOT, BOTH MONKEY AND FOOD FALL TO HERE

BOTH MONKEY AND FOOD FALL TO HERE FOR MEDIUM-SPEED SHOT

PLACE WHERE BOTH MEET FOR LOW-SPEED SHOT

ANSWER:

WHEN THE MONKEY REMAINS AT REST, THE KEEPER SHOULD AIM ABOVE THE MONKEY TO COMPENSATE FOR GRAVITY. BUT WHEN THE MONKEY DROPS, THE KEEPER SHOULD AIM DIRECTLY AT THE MONKEY. BECAUSE OF GRAVITY, THE FOOD (LIKE ANY PROJECTILE) WILL FALL BELOW THIS DIRECT STRAIGHT LINE. HOW FAR BELOW? AS FAR BELOW AS THE MONKEY FALLS IN THE SAME TIME. SO THE MONKEY AND THE FOOD WILL FALL THE SAME VERTICAL DISTANCE IN THE SAME TIME, AND MID-AIR CONTACT IS MADE.

CONCEPTUAL Physics

When the ball at the end of the string swings to its lowest point, the string is cut by a sharp razor.

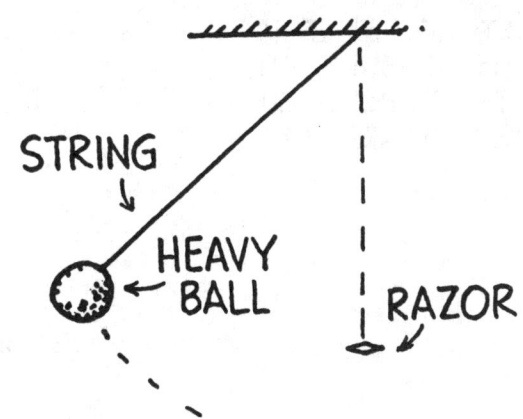

Which path will the ball then follow?

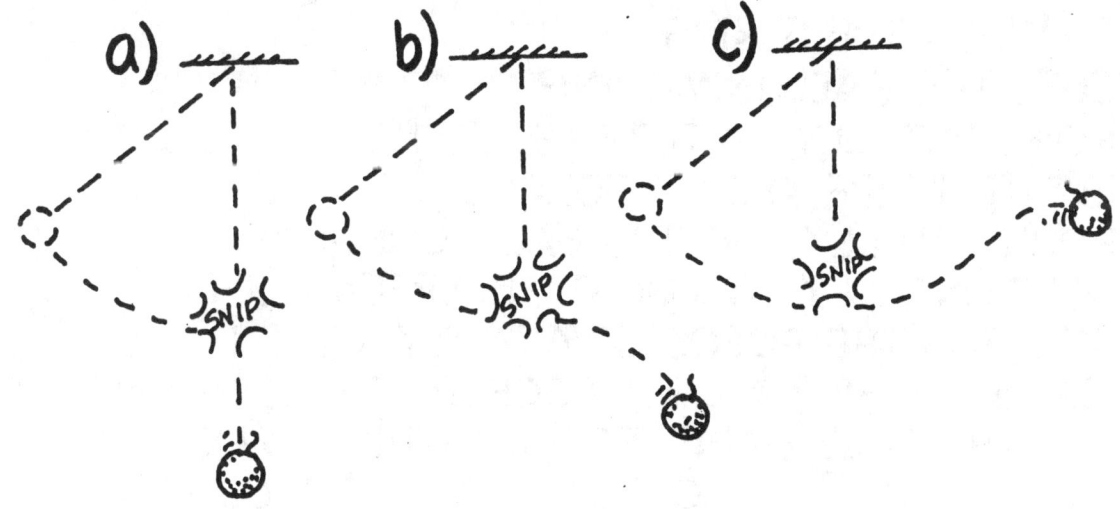

CONCEPTUAL Physics

WHEN THE BALL AT THE END OF THE STRING SWINGS TO ITS LOWEST POINT, THE STRING IS CUT BY A SHARP RAZOR.

WHICH PATH WILL THE BALL THEN FOLLOW?

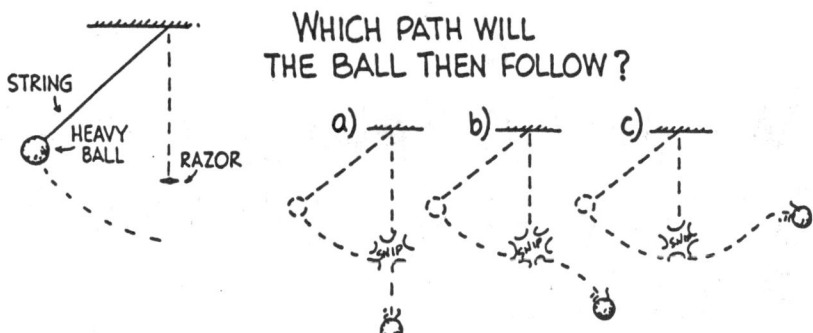

ANSWER:

WHEN THE STRING IS CUT, THE BALL IS MOVING HORIZONTALLY. AFTER THE STRING IS CUT THERE ARE NO FORCES HORIZONTALLY, SO THE BALL CONTINUES HORIZONTALLY AT CONSTANT SPEED. BUT THERE IS THE FORCE OF GRAVITY WHICH CAUSES THE BALL TO ACCELERATE DOWNWARD, SO THE BALL GAINS SPEED IN THE DOWNWARD DIRECTION. THE COMBINATION OF A CONSTANT HORIZONTAL SPEED AND A DOWNWARD GAIN IN SPEED PRODUCES THE CURVED PATH CALLED A PARABOLA. THE BALL CONTINUES ALONG PATH b --- A PARABOLIC PATH.

CHAP. 4

CONCEPTUAL Physics

WHEN THE PELLET FIRED INTO THE SPIRAL TUBE EMERGES, WHICH PATH WILL IT FOLLOW?
(NEGLECT GRAVITY)

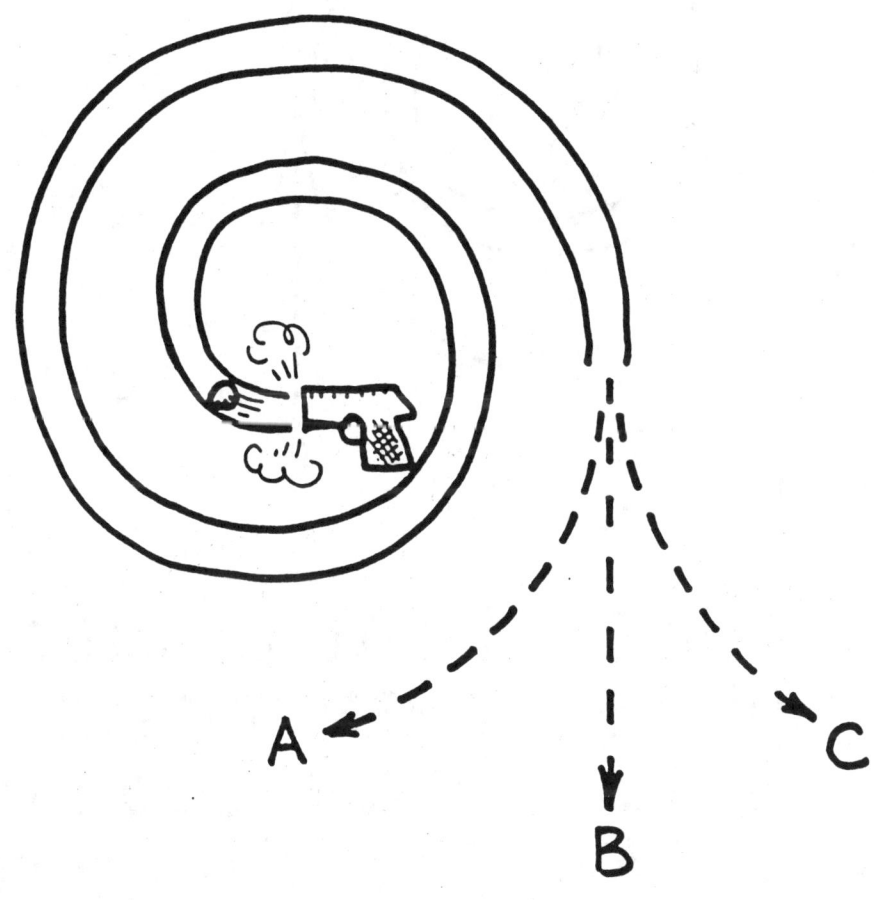

CONCEPTUAL Physics

When the pellet fired into the spiral tube emerges, which path will it follow? (Neglect gravity)

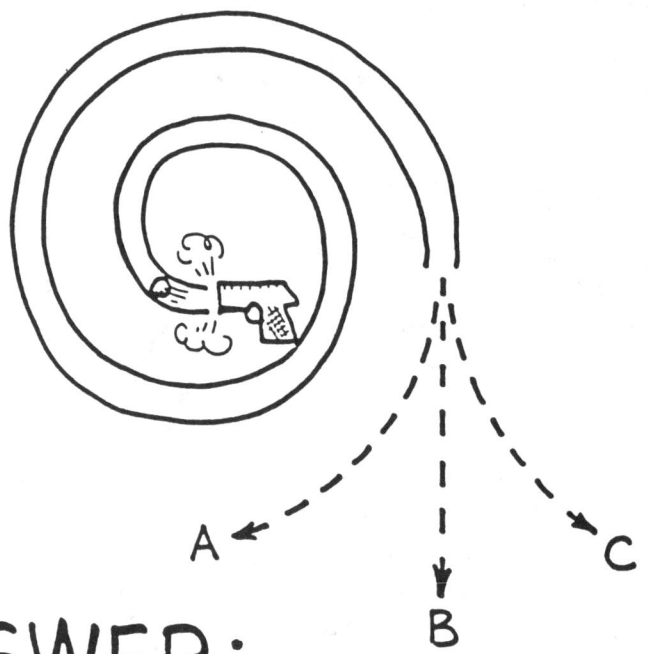

ANSWER:

While in the tube it is forced to curve, but when it gets outside, no force is exerted on the pellet and (law of inertia) it follows a straight-line path... B!

CONCEPTUAL Physics

WHICH ENCOUNTERS THE GREATER FORCE OF AIR RESISTANCE --- A FALLING ELEPHANT OR A FALLING FEATHER?

ANSWER:

There is a greater force of air resistance on the falling elephant, which "plows through" more air than the feather in getting to the ground. The elephant encounters several newtons of air resistance, which compared to its huge weight has practically no effect on its rate of fall. Only a small fraction of a newton acts on the feather, but the effect is significant because the feather weighs only a fraction of a newton.

Remember to distinguish between a force itself and the effect it produces!

CONCEPTUAL Physics

Two smooth balls of exactly the same size, one made of wood and the other of iron, are dropped from a high building to the ground below. The ball to encounter the greater force of air resistance on the way down is the

a) wooden ball

b) iron ball

c) ... both the same

CHAP. 4

CONCEPTUAL **Physics**

Two smooth balls of exactly the same size, one made of wood and the other of iron, are dropped from a high building to the ground below. The ball to encounter the greater force of air resistance on the way down is the

a) wooden ball

b) iron ball

c) ... both the same

The answer is b:

Air resistance depends on both the size and speed of a falling object. Both balls have the same size, but the heavier iron ball falls faster through the air and encounters more air resistance in its fall.

Be careful to distinguish between the *amount* of air drag and the *effect* of that air drag. If the greater air drag on the faster ball is small compared to the weight of the ball, it won't be very effective in reducing acceleration. Like 2 newtons of air drag on a 20-newton ball has less effect on fall than 1 newton of air drag on a 2-newton ball.

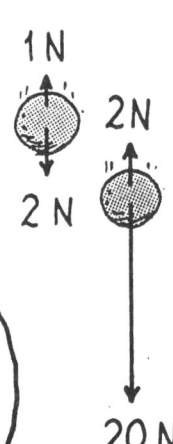

CHAP. 4

CONCEPTUAL **Physics**

As she falls faster and faster through the <u>air</u>, her acceleration
  a) increases
  b) decreases
  c) remains the same

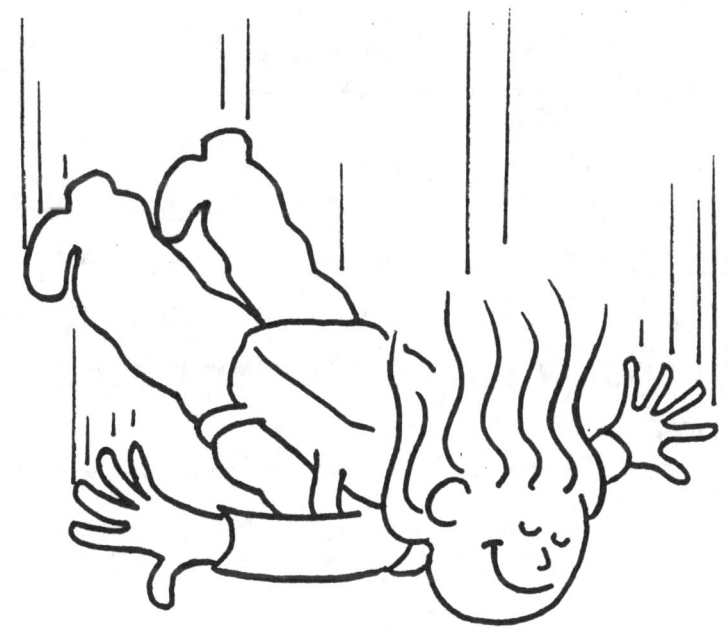

CHAP. 4

CONCEPTUAL Physics

As she falls faster and faster through the <u>air</u>, her acceleration
a) increases
b) decreases
c) remains the same

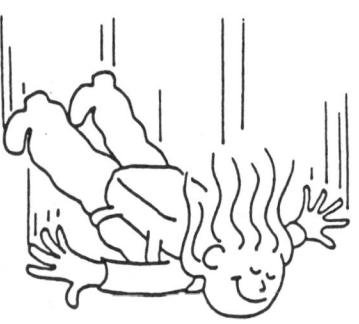

The answer is b:

Acceleration decreases because the net force on her decreases. Net force is equal to her weight minus her air resistance, and since air resistance increases with increasing speed, net force and hence acceleration decrease. By Newton's 2nd law:

$$a = \frac{F_{NET}}{m} = \frac{(mg - R)}{m}$$

where mg is her weight, and R is the air resistance she encounters. As R increases, a decreases. Note that if she falls fast enough so that $R = mg$, $a = 0$, then with no acceleration she falls at constant velocity.

> Go an extra step in the equation for Newton's 2nd law (divide mg and R by m) and get
> $$a = g - \frac{R}{m}$$
> Note that the acceleration a will always be less than g if air resistance R impedes falling. Only when $R = 0$ does $a = g$.

CONCEPTUAL Physics

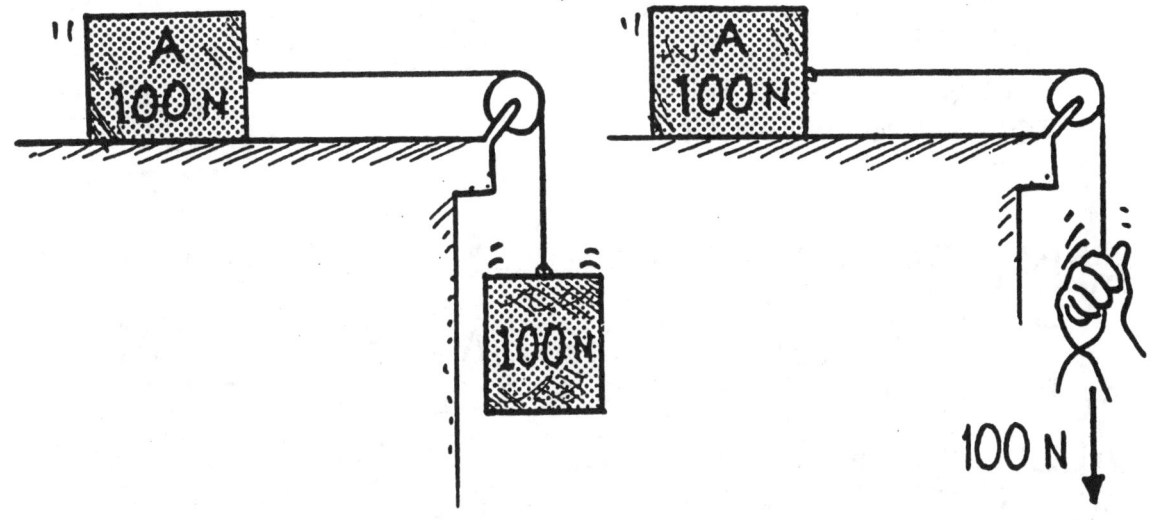

In both cases an applied force of 100 N accelerates the 100-N block.

In which case is the acceleration greater?

CONCEPTUAL Physics

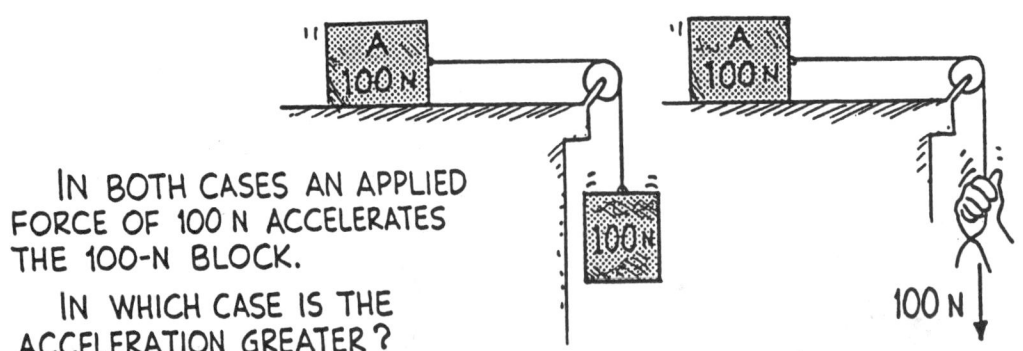

IN BOTH CASES AN APPLIED FORCE OF 100 N ACCELERATES THE 100-N BLOCK.

IN WHICH CASE IS THE ACCELERATION GREATER?

ANSWER:

THE ONE-BLOCK SYSTEM HAS THE GREATER ACCELERATION. THIS IS BECAUSE DIFFERENT ACCELERATIONS ARE PRODUCED WHEN THE SAME FORCE IS APPLIED TO SYSTEMS OF DIFFERENT MASS. TWICE THE MASS IS BEING ACCELERATED IN THE TWO-BLOCK SYSTEM, SO ITS ACCELERATION IS HALF THAT OF THE ONE-BLOCK SYSTEM.

(CAN YOU SEE THAT SINCE THE APPLIED FORCE EQUALS THE WEIGHT OF THE ONE-BLOCK SYSTEM, IT ACCELERATES AT g? AND THAT THE TWO-BLOCK SYSTEM ACCELERATES AT $g/2$? AND CAN YOU SEE THAT THE ROPE TENSIONS IN THE TWO CASES ARE UNEQUAL? THAT IT MUST BE 50 N FOR THE TWO-BLOCK SYSTEM?)

CONCEPTUAL Physics

For every force there exists an equal and opposite force. Consider action and reaction forces in the case of a rock falling under the influence of gravity. If action is considered to be that of the earth pulling down on the rock, can you clearly identify the reaction force?

CONCEPTUAL Physics

For every force there exists an equal and opposite force. Consider action and reaction forces in the case of a rock falling under the influence of gravity. If action is considered to be that of the earth pulling down on the rock, can you clearly identify the reaction force?

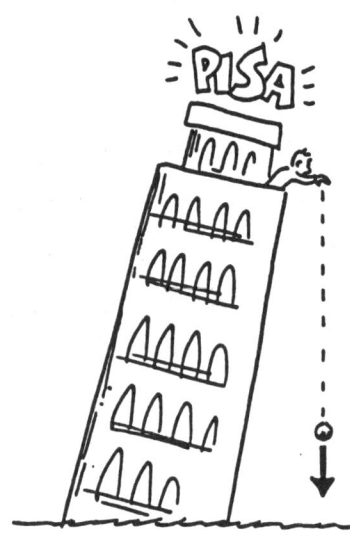

ANSWER:

The recipe for action-reaction forces is simple enough: if **A** exerts force on **B**, then in turn, **B** exerts force on **A**.

So if action is the earth pulling down on the falling rock, reaction is simply the falling rock pulling up on the earth. Does this mean that the acceleration of the rock and the earth should be the same? Not at all, but only because the earth's mass is so much greater than that of the falling rock.

CHAP. 4

CONCEPTUAL Physics

IF A MACK TRUCK AND A VOLKSWAGEN HAVE A HEAD-ON COLLISION, WHICH VEHICLE WILL EXPERIENCE THE GREATER IMPACT FORCE?

a) THE MACK TRUCK
b) THE VOLKSWAGEN
c) BOTH THE SAME
d) ... IT DEPENDS ON OTHER FACTORS

CONCEPTUAL Physics

IF A MACK TRUCK AND A VOLKSWAGEN HAVE A HEAD-ON COLLISION, WHICH VEHICLE WILL EXPERIENCE THE GREATER IMPACT FORCE?

a) THE MACK TRUCK
b) THE VOLKSWAGEN
c) BOTH THE SAME
d) ...IT DEPENDS ON OTHER FACTORS

THE ANSWER IS C:

BOTH WILL EXPERIENCE THE SAME IMPACT FORCE, IN ACCORD WITH NEWTON'S 3rd LAW. THE FORCE THAT BODY **A** EXERTS ON BODY **B** IS EQUAL AND OPPOSITE TO THE FORCE THAT BODY **B** EXERTS ON BODY **A**. THE *EFFECTS* OF THESE FORCES, HOWEVER, ARE QUITE DIFFERENT --- NOTE FROM NEWTON'S 2nd LAW

$$\frac{F_{TRUCK}}{m_{TRUCK}} = a \; ; \; \frac{F_{CAR}}{m_{CAR}} = a$$

SO THE CAR DECELERATES MUCH MORE THAN THE MASSIVE TRUCK.

CHAP. 4

CONCEPTUAL Physics

WHAT WILL BE THE ACCELERATION OF A ROCK THROWN STRAIGHT UPWARD AT THE MOMENT IT REACHES THE TIPPITY-TOP OF ITS TRAJECTORY?

CHAP. 4

CONCEPTUAL Physics

WHAT WILL BE THE ACCELERATION OF A ROCK THROWN STRAIGHT UPWARD AT THE MOMENT IT REACHES THE TIPPITY-TOP OF ITS TRAJECTORY?

ANSWER:

ALTHOUGH ITS SPEED AND VELOCITY AT THE TOP WILL BOTH INSTANTANEOUSLY BE ZERO, ITS ACCELERATION WILL BE g, OR 9.8 m/s^2. REMEMBER, ACCELERATION IS NOT SPEED OR VELOCITY --- IT IS THE *RATE* AT WHICH VELOCITY CHANGES. A MOMENT BEFORE OR AFTER THE ROCK REACHES THE TOP, IT IS MOVING, WHICH IS EVIDENCE THAT ITS VELOCITY IS CHANGING AT EVERY INSTANT. THE ROCK UNDERGOES A CHANGE AS IT PASSES THROUGH THE ZERO VALUE OF VELOCITY JUST AS IT UNDERGOES THE SAME RATE OF CHANGE PASSING THROUGH ANY OTHER VALUE OF VELOCITY.

OR LOOK AT IT VIA NEWTON'S 2ND LAW. AT THE TOP OR ANYWHERE IN ITS PATH, THE ROCK HAS BOTH WEIGHT AND MASS, AND

$$a = \frac{F}{m} = \frac{\cancel{m}g}{\cancel{m}} = g.$$

CONCEPTUAL Physics

A 1-kg ROCK IS THROWN AT 10 m/s STRAIGHT UPWARD. NEGLECTING AIR RESISTANCE, WHAT IS THE NET FORCE THAT ACTS ON IT WHEN IT IS HALF WAY TO THE TOP OF ITS PATH?

CHAP. 4

CONCEPTUAL Physics

A 1-kg ROCK IS THROWN AT 10 m/s STRAIGHT UPWARD. NEGLECTING AIR RESISTANCE, WHAT IS THE NET FORCE THAT ACTS ON IT WHEN IT IS HALF WAY TO THE TOP OF ITS PATH?

ANSWER:

IN THE ABSENCE OF AIR RESISTANCE, THE ONLY FORCE EXERTED ON THE 1 kg ROCK IS SIMPLY THE FORCE OF GRAVITY --- mg. THAT'S 9.8 NEWTONS, AT <u>ANY</u> SPEED AND ANYWHERE ALONG ITS TRAJECTORY!

NET FORCE = mg
= (1 kg)(9.8 m/s²)
= 9.8 N

DON'T CONFUSE FORCE VECTORS WITH VELOCITY VECTORS!

WHAT'S THE ACCELERATION OF THE ROCK AT THE TOP OF ITS PATH?

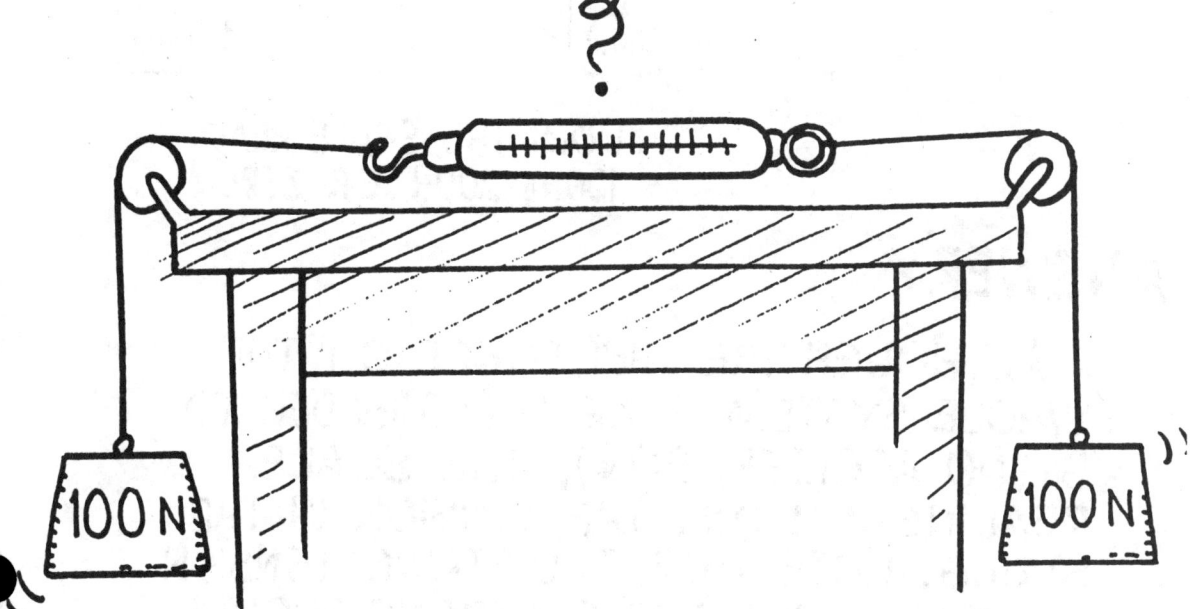

CONCEPTUAL Physics

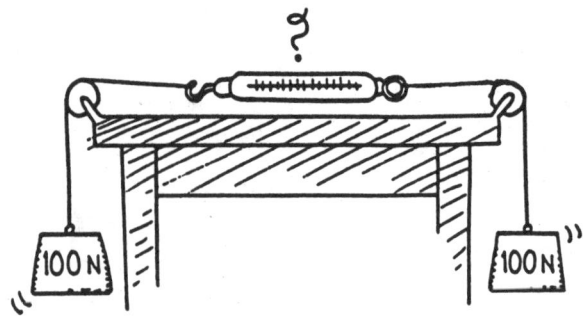

DOES THE SCALE READ 100N, 200N, OR ZERO?

ANSWER:

ALTHOUGH THE NET FORCE ON THE WHOLE SYSTEM IS ZERO (AS EVIDENCED BY NO ACCELERATION), THE SCALE READING IS 100N, THE TENSION IN THE STRING. NOTE THAT THE STRING TENSION IS 100N IN ALL THE POSITIONS SHOWN.

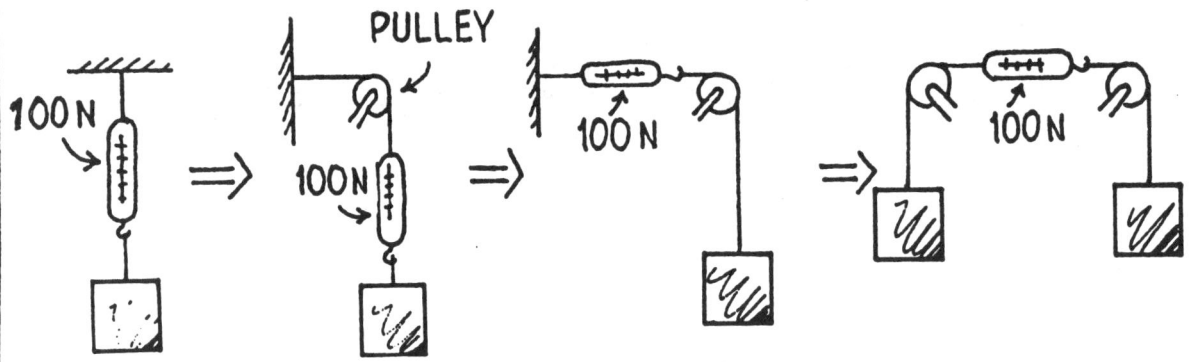

CHAP. 4

CONCEPTUAL **Physics**

Arnold Strongman and Suzie Small pull on opposite ends of a rope in a tug of war. The greatest force exerted on the rope is by

a) Arnold

b) Suzie

c) ... both the same

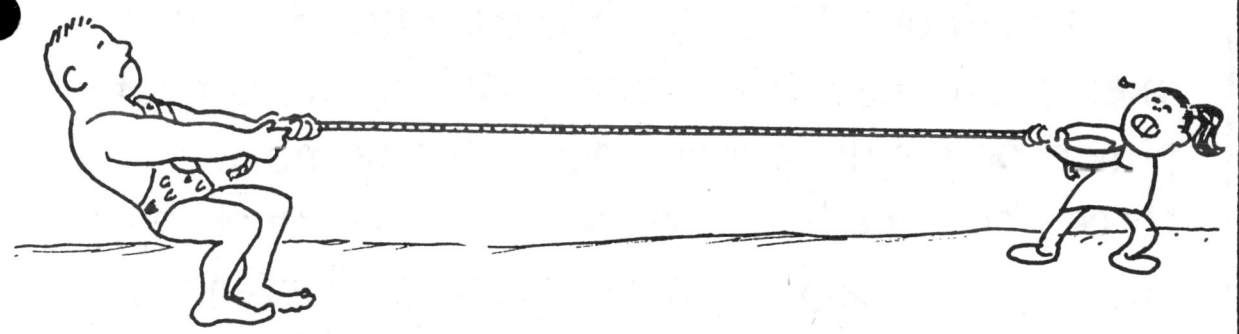

Assume the rope's mass is negligible.

CHAP. 4

CONCEPTUAL Physics

Arnold Strongman and Suzie Small pull on opposite ends of a rope in a tug of war. The greatest force exerted on the rope is by

a) Arnold
b) Suzie
c) ...both the same

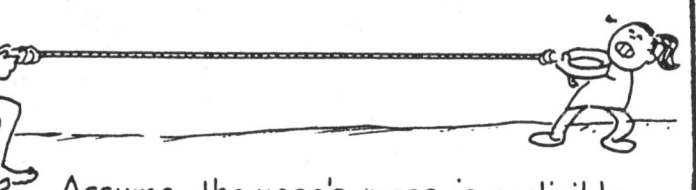

Assume the rope's mass is negligible.

The answer is C:

Arnold can pull no harder on the rope than Suzie. Rope tension is the same all along the rope, including the ends. Just as a wheel on ice can exert no more force on the ice than the ice exerts on the wheel, and just as one cannot punch an empty paper bag with any more force than the bag can exert on the puncher, Arnold can exert no more force on his end of the rope than Suzie exerts on her end.

Arnold can push harder on the ground than Suzie can, so even though the pulls on the rope are the same, Arnold will likely win the tug of war!

CONCEPTUAL Physics

Two identical rubber bands connect masses A and B to a string over a frictionless pulley of negligible mass. The amount of stretch is greater in the band that connects

a) A

b) B

c) Both the same

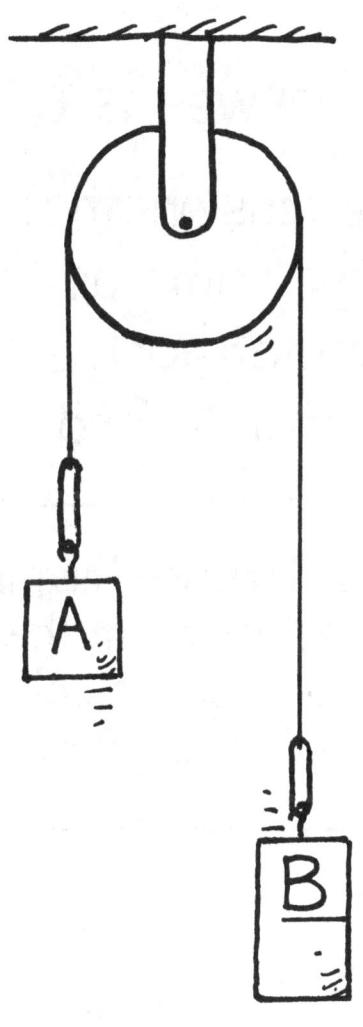

CHAP. 4

CONCEPTUAL Physics

Two identical rubber bands connect masses A and B to a string over a frictionless pulley of negligible mass. The amount of stretch is greater in the band that connects

a) A
b) B
c) Both the same

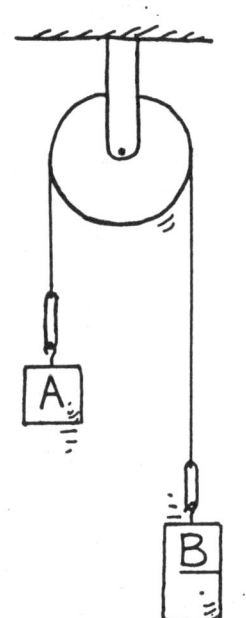

The answer is c:

The tension that stretches the rubber bands is the same as the tension in the string — same at both ends, in accord with Newton's 3rd law.

To better see this, imagine the rubber bands are farther from the ends of the string. If the tension all along the string is the same, likewise for the rubber bands.

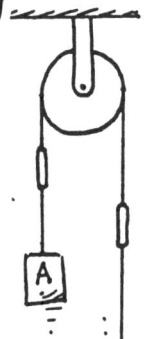

CHAP. 4

CONCEPTUAL Physics

The brakes are slammed on a speeding truck and it skids to a stop. If the truck were heavily loaded so it had twice the total mass, the skidding distance would be

a) the same
b) 1½ times as far
c) twice as far
d) four times as far

CONCEPTUAL Physics

The brakes are slammed on a speeding truck and it skids to a stop. If the truck were heavily loaded so it had twice the total mass, the skidding distance would be

a) the same
b) 1½ times as far
c) twice as far
d) four times as far

Answer:

Twice the mass means the skidding tires will bear against the road with twice the force, which results in twice the friction. Twice as much friction acting on twice as much mass produces the same deceleration and hence the same stopping distance.

Twice the *speed* would produce four times the stopping distance.

CONCEPTUAL Physics

When the spool is pulled horizontally to the right, it will roll toward the

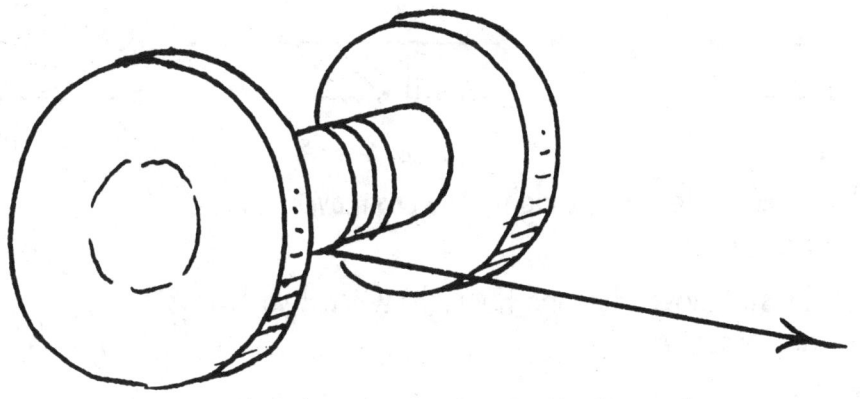

a) right
b) left

CHAP. 4

CONCEPTUAL Physics

When the spool is pulled horizontally to the right, it will roll toward the

a) right
b) left

The answer is a:

The spool rolls toward the right. To see this, consider the peg being pulled by the string in the sketches below:

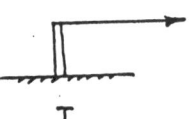

I   II   III

Sketch I: When pulled to the right, the peg moves toward the right.

Sketch II: It still moves toward the right when two "wings" are attached.

Sketch III: When the top is capped the peg still moves to the right.

We can go further: The spool rolls because the applied force produces a torque on the spool. If the string is pulled at a slight angle above the horizontal, the torque still rolls it to the right. Beyond a steeper angle, however, the spool will roll to the left. What angle? At the angle where the line of action of the string intersects the point where the spool makes contact with the surface.

No torque acts on the spool when the cosine of the angle equals r/R. For smaller angles a torque is produced that makes it roll to the right; for larger angles, an opposite torque is produced that rolls it to the left.

The smaller the radius of the spindle compared to the outer radius, the higher the string must be pulled to make the spool roll in the opposite direction.

CONCEPTUAL Physics

Jocko, who has a mass of 60 kg and stands at rest on ice, catches a 20 kg ball that is thrown to him at 10 km/h. How fast does Jocko and the ball move across the ice?

CONCEPTUAL Physics

JOCKO, WHO HAS A MASS OF 60 kg AND STANDS AT REST ON ICE, CATCHES A 20 kg BALL THAT IS THROWN TO HIM AT 10 km/h. HOW FAST DOES JOCKO AND THE BALL MOVE ACROSS THE ICE?

ANSWER:

THE MOMENTUM BEFORE THE CATCH IS ALL IN THE BALL, 20 kg × 10 km/h = 200 kg·km/h. THIS IS ALSO THE MOMENTUM AFTER THE CATCH, WHERE THE MOVING MASS IS 80 kg --- 60 kg FOR JOCKO AND 20 kg FOR THE CAUGHT BALL.

$$80 \text{ kg} \times v = 200 \text{ kg·km/h}$$

$$v = \frac{200 \text{ kg·km/h}}{80 \text{ kg}} = 2.5 \text{ km/h}$$

CHAP. 5

CONCEPTUAL Physics

Which would be more damaging; driving into a massive concrete wall, or driving at the same speed into a head-on collision with an identical car traveling toward you at the same speed?

CHAP. 5

CONCEPTUAL Physics

Which would be more damaging; driving into a massive concrete wall, or driving at the same speed into a head-on collision with an identical car traveling toward you at the same speed?

ANSWER: Both cases are equivalent, because either way, your car rapidly decelerates to a dead stop. The dead stop is easy to see when hitting the wall, and a little thought will show the same is true when hitting the car. If the oncoming car were traveling slower, with less momentum, you'd keep going after collision with more "give," and less damage (to you!). But if the oncoming car had more momentum than you, IT would keep going and you'd snap into a sudden reverse with greater damage. Identical cars at equal speeds means equal momenta --- zero before, zero after collision.

CONCEPTUAL Physics

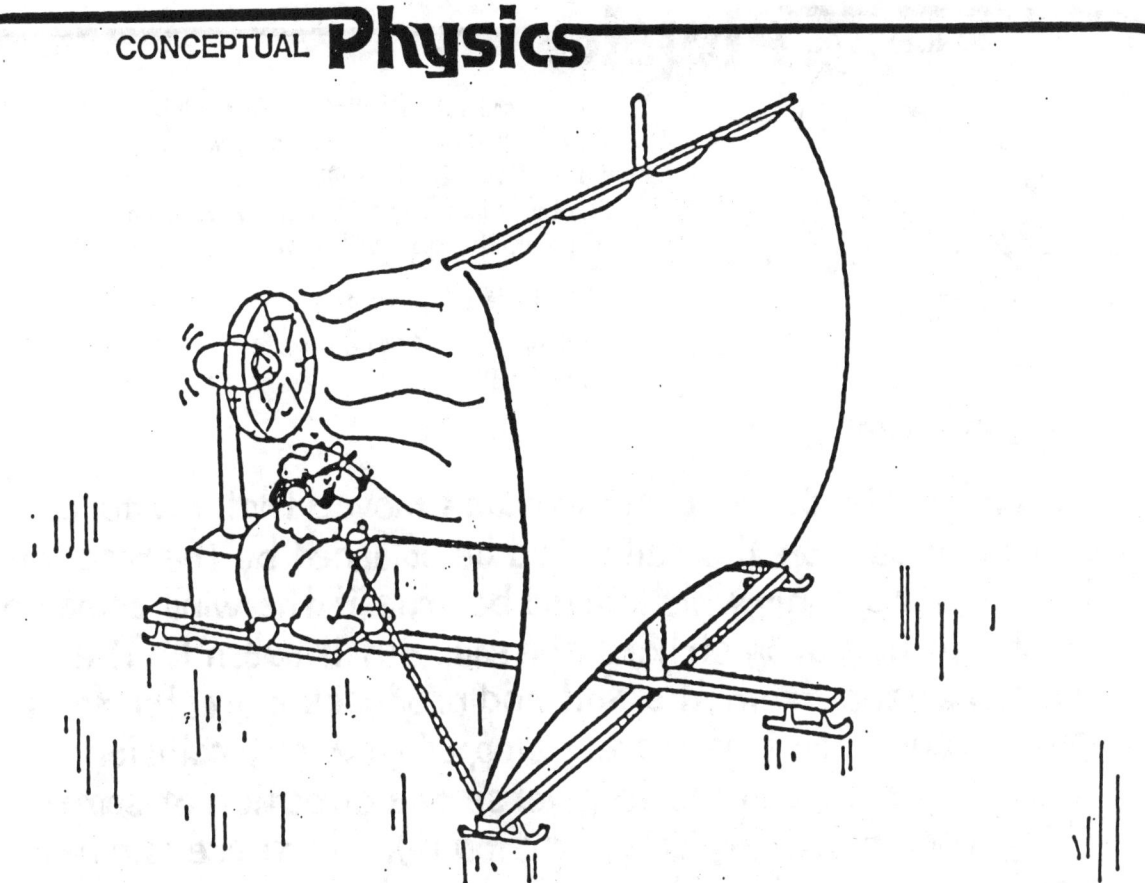

An ice sailcraft is stalled on a frozen lake on a windless day. A large fan blows air into the sail. If all the wind produced by the fan strikes and bounces backward from the sail, the craft will move

a) to the left (backward)
b) to the right (forward)
c) not at all

CONCEPTUAL Physics

An ice sailcraft is stalled on a frozen lake on a windless day. A large fan blows air into the sail. If all the wind produced by the fan strikes and bounces backward from the sail, the craft will move

a) to the left (backward)
b) to the right (forward)
c) not at all

The answer is b:

You might think the craft wouldn't move -- that the force of wind impact on the sail would be balanced by the reaction force on the fan -- which would be true if the wind came to an abrupt halt upon striking the sail. But it doesn't. The wind bounces from the sail and produces a greater force on the sail than if it merely stopped (like any collision, more force is required to reverse the direction of something than to merely start or stop it). So there is a net force on the craft and a forward acceleration.

Or consider impulse and momentum. The impulse on the sail is greater than the impulse on the fan. Why? Because the air undergoes more change in momentum bouncing from the sail than starting from the fan.

Note there are two force pairs to consider: (1) the fan-air force pair, and (2) the air-sail force pair. Because of bouncing, the air-sail pair is greater. Solid vectors show forces exerted on the craft; dashed vectors show forces exerted on the air. The net force on the craft is forward, to the right.

Why not simply turn the fan around and omit the sail?

CHAP. 5

CONCEPTUAL Physics

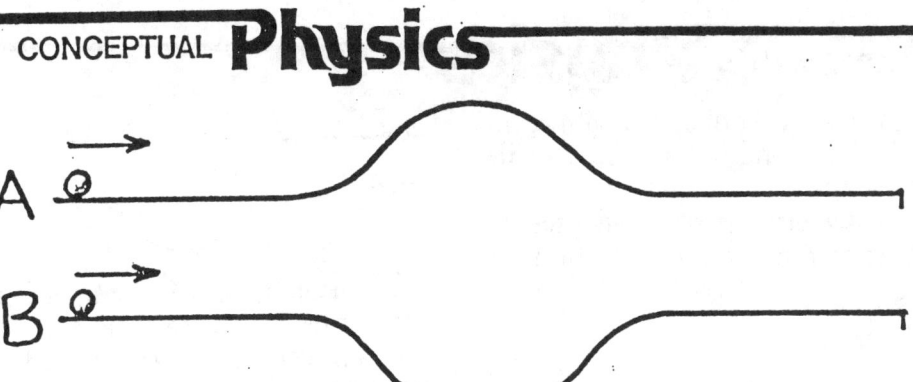

Two smooth tracks of equal length have "bumps" — A up, and B down, both of the same curvature. If two balls start simultaneously with the same initial speed, the ball to complete the journey first is along

    a) Track A

    b) Track B

    c) ... both take the same time

If the initial speed = 2 m/s, and the speed of the ball at the bottom of the curve on track B is 3 m/s, then the speed of the ball at the top of the curve on Track A is

    d) 1 m/s   e) >1 m/s   f) <1 m/s

CONCEPTUAL Physics

Two smooth tracks of equal length have "bumps"— A up, and B down, both of the same curvature. If two balls start simultaneously with the same initial speed, the ball to complete the journey first is along

    a) Track A
    b) Track B
    c) ... both take the same time

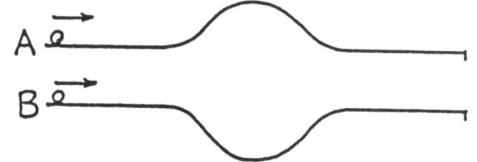

If the initial speed = 2 m/s, and the speed of the ball at the bottom of the curve on track B is 3 m/s, then the speed of the ball at the top of the curve on Track A is

    d) 1 m/s   e) >1 m/s   f) <1 m/s

The answers are **b** and **f**:

Although both balls have the same speed on the level parts of the tracks, the speeds along the curved parts differ. The speed of the ball everywhere along curve B is greater than the initial speed, whereas everywhere along curve A it is less. So the ball on track B finishes first.

Does the gain in speed at B's bottom equal the loss at A's top? No! Speed isn't conserved: *energy* is. The loss in *kinetic energy* at the top of A will be equal to the gain in *kinetic energy* at the bottom of B — if there is enough energy to begin with.

There isn't, because the initial KE [$\frac{1}{2}m 2^2$] is less than the *gain* in KE at the bottom of B [$\frac{1}{2}m(3^2-2^2)$]. At 2 m/s, the ball will not even make it to the top of A's curve.

CHAP. 6

CONCEPTUAL Physics

FOR THE SAME FORCE, WHY IS THE SPEED OF A CANNONBALL GREATER WHEN SHOT FROM A CANNON WITH A LONGER BARREL?

ANSWER:

THERE ARE TWO MAIN REASONS FOR THE GREATER SPEED. A CANNONBALL WITH GREATER SPEED HAS GREATER MOMENTUM AND GREATER KINETIC ENERGY. HOW DOES IT GET GREATER MOMENTUM FOR THE SAME APPLIED FORCE? BY A GREATER IMPULSE, WHICH IS "FORCE × TIME." THE TIME DURING WHICH THE FORCE ACTS IS LONGER IN THE LONG BARREL! OR HOW DOES THE CANNONBALL GET MORE KINETIC ENERGY? BY GREATER WORK, WHICH IS "FORCE × DISTANCE." THE GREATER DISTANCE THE FORCE ACTS IN THE BARREL PRODUCES MORE WORK = MORE KINETIC ENERGY!

CONCEPTUAL Physics

Three baseballs are thrown from the top of the cliff along paths A, B and C. If their initial speeds are the same and there is no air resistance, the ball that strikes the ground below with the greatest speed will follow path

a) A   b) B   c) C   d) either A or C

e) all strike with the same speed

HINT: Consider energy conservation

CHAP. 6

CONCEPTUAL Physics

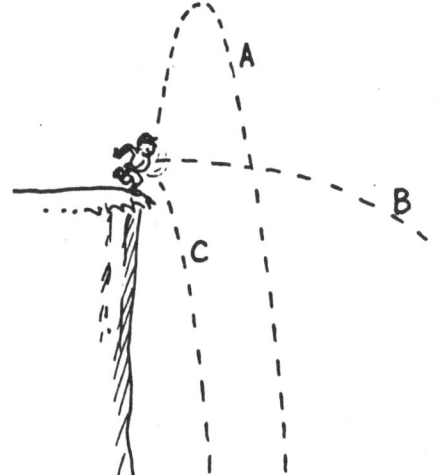

Three baseballs are thrown from the top of the cliff along paths A, B and C. If their initial speeds are the same and there is no air resistance, the ball that strikes the ground below with the greatest speed will follow path

a) A   b) B   c) C   d) either A or C
e) all strike with the same speed

The answer is e; the speed of impact for each ball is the same. With respect to the ground below, the initial kinetic + potential energy of each ball is the same. This amount of energy becomes the kinetic energy at impact. So for equal masses, equal kinetic energies means the same speed.

$$(mgh + \tfrac{1}{2}mv^2)_{initial} = (\tfrac{1}{2}mU^2)_{final}$$

CHAP. 6

CONCEPTUAL Physics

A pair of upright meter sticks, with their lower ends against a wall, are allowed to fall to the floor. One is bare, and the other has a heavy weight attached to its upper end. The stick to hit the floor first is the

a) bare stick
b) weighted stick
c) ...both the same

Try it and see!

CHAP. 7

CONCEPTUAL Physics

A pair of upright meter sticks, with their lower ends against a wall, are allowed to fall to the floor. One is bare, and the other has a heavy weight attached to its upper end. The stick to hit the floor first is the

a) bare stick

b) weighted stick

c) ...both the same

Try it and see!

The answer is a:

In falling, both sticks rotate about an axis at the lower end where the wall and floor meet. Their rate of rotation depends on their rotational inertias. The stick with the heavy weight at its upper end has more rotational inertia and is more lazy in rotating about its lower end. So the bare stick rotates to the floor in the shortest time.

Just as important is the *torque* that acts on each stick. Even though the weighted stick has a greater torque, it's not enough greater to change the outcome. But that's another story.

CHAP. 7

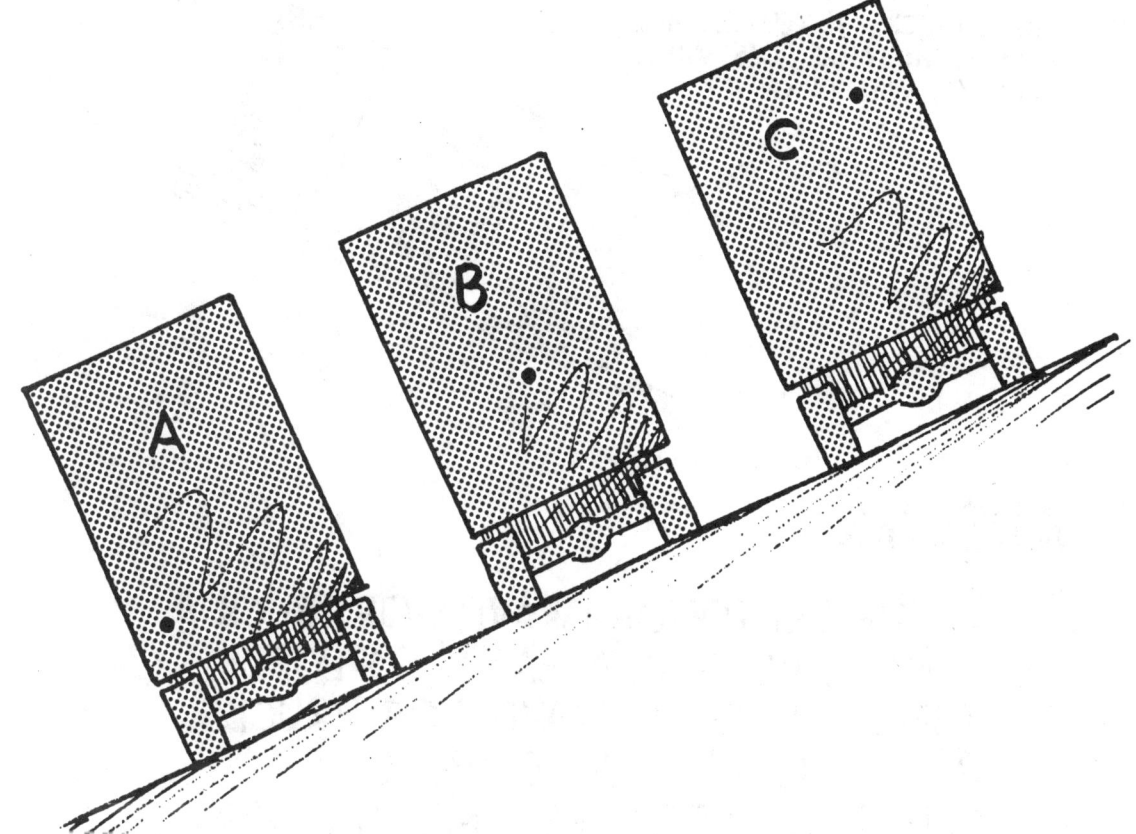

CONCEPTUAL Physics

THE CENTERS OF GRAVITY OF THREE TRUCKS PARKED ON A HILL ARE SHOWN BY THE DOTS. WHICH TRUCK(S) WILL TIP OVER?

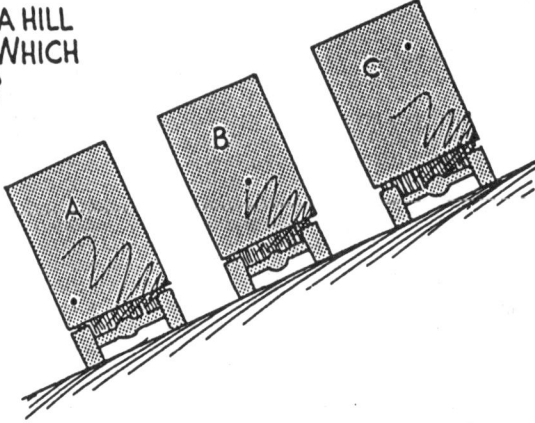

ANSWER:

THE CENTER OF GRAVITY OF TRUCK **A** IS NOT ABOVE AN AREA OF SUPPORT; THE CENTERS OF GRAVITY OF TRUCKS **B** AND **C** ARE ABOVE AREAS OF SUPPORT. THEREFORE ONLY TRUCK **A** WILL TIP OVER.

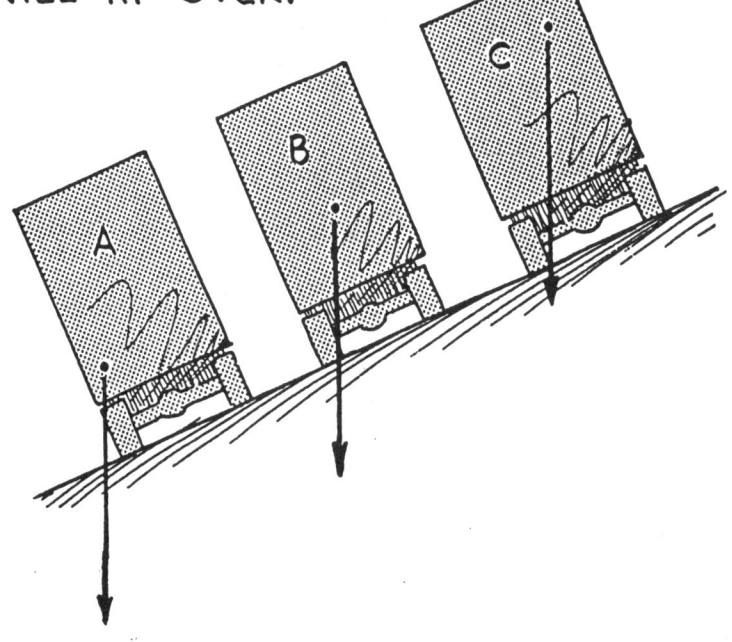

CONCEPTUAL Physics

If a small object is placed on a rotating disk (like a record player) it will slide off the edge. Suppose you fasten a bar on the disk as shown in the top view, and allow the object to slide from an inside position, along the bar, and off the edge. Which of the paths shown would be most likely?

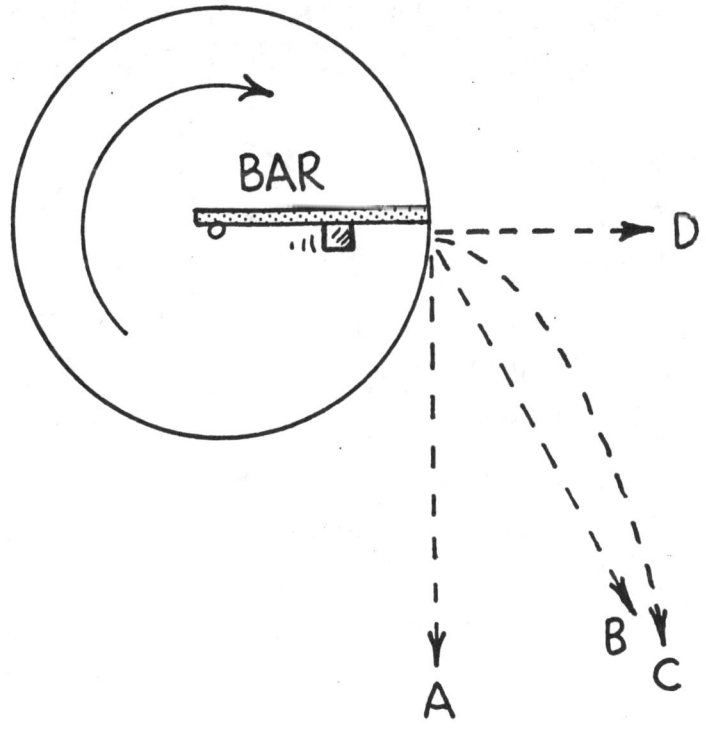

CONCEPTUAL Physics

IF A SMALL OBJECT IS PLACED ON A ROTATING DISK (LIKE A RECORD PLAYER) IT WILL SLIDE OFF THE EDGE. SUPPOSE YOU FASTEN A BAR ON THE DISK AS SHOWN IN THE TOP VIEW, AND ALLOW THE OBJECT TO SLIDE FROM AN INSIDE POSITION, ALONG THE BAR, AND OFF THE EDGE. WHICH OF THE PATHS SHOWN WOULD BE MOST LIKELY?

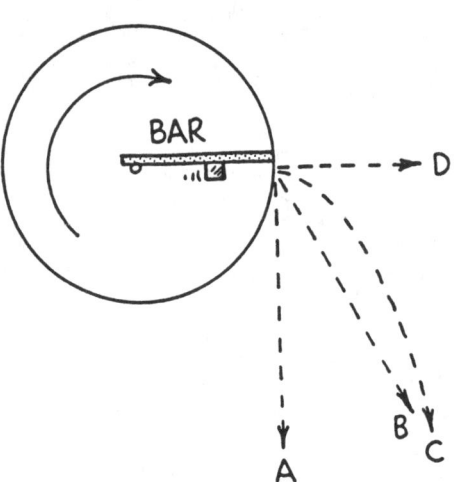

ANSWER:

PATH **B** IS THE LIKELY PATH. PATH **A** WOULD BE THE CASE IF THE OBJECT HAD NO RADIAL COMPONENT OF MOTION AND SIMPLY FLEW OFF THE EDGE. PATH **D** WOULD OCCUR IF THE DISK WEREN'T SPINNING AND THE OBJECT SLID RADIALLY OUTWARD. PATH **B** IS SIMPLY THE RESULTANT OF PATH **A** (TANGENTIAL COMPONENT) AND PATH **D** (RADIAL COMPONENT). PATH **C** CURVES IN VIOLATION OF THE LAW OF INERTIA.

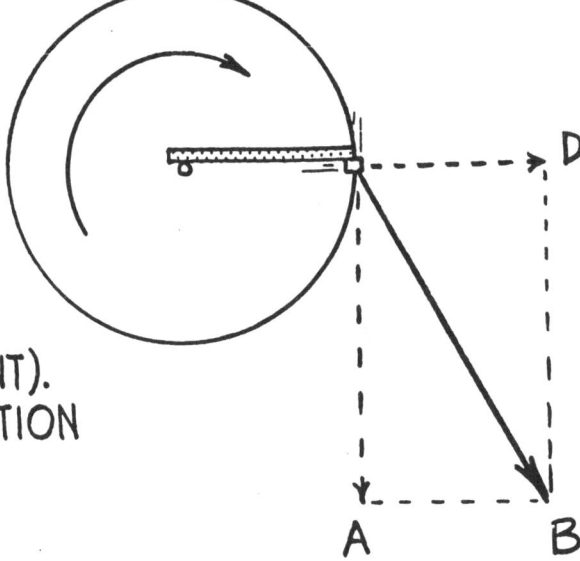

CHAP. 7

CONCEPTUAL Physics

James finds it difficult to muster enough torque to turn the stubborn bolt with the wrench. He wishes he had a pipe handy to effectively lengthen the wrench handle, but doesn't. He does, however, have a piece of rope. Will torque be increased if he pulls as hard on the rope as shown?

CHAP. 7

CONCEPTUAL Physics

James finds it difficult to muster enough torque to turn the stubborn bolt with the wrench. He wishes he had a pipe handy to effectively lengthen the wrench handle, but doesn't. He does, however, have a piece of rope. Will torque be increased if he pulls as hard on the rope as shown?

ANSWER:

No, the torque will be the same because the lever-arm distance is the same in both cases. The lever arm is not the distance between axis of turning and the point of application of the force, but the distance from the turning axis to the "line of action" of the applied force. Note the line of action, and hence the lever arm is the same in both cases.

The pipe that extends the length of the wrench handle puts the line of action farther from the turning axis -- the rope does not.

CHAP. 7

CONCEPTUAL Physics

Which will roll down a hill faster, a can of regular fruit juice or a can of frozen fruit juice?

a) regular fruit juice

b) frozen fruit juice

c) depends on the relative sizes and weights of the cans

CHAP. 7

CONCEPTUAL Physics

Which will roll down a hill faster, a can of regular fruit juice or a can of frozen fruit juice?

a) regular fruit juice
b) frozen fruit juice
c) depends on the relative sizes and weights of the cans

The answer is a:

The regular fruit juice has an appreciably greater acceleration down an incline than the can of frozen juice. Why? Because the regular juice is a liquid and is not made to roll with the can, as the solid juice does. Most of the liquid effectively slides down the incline inside the rolling can. The can of liquid therefore has very little rotational inertia compared to its mass. The solid juice, on the other hand, is made to rotate, giving the can more rotational inertia.

Any freely sliding object will beat any rotating object on the same incline because none of its potential energy is given to rotational kinetic energy.

CHAP. 7

CONCEPTUAL Physics

When she shakes the basket full of berries, the larger berries will

a) sink to the bottom

b) go to the top

c) not particularly sink nor rise, but like the smaller berries, be randomly distributed

CHAP. 7

CONCEPTUAL Physics

When she shakes the basket full of berries, the larger berries will

a) sink to the bottom
b) go to the top
c) not particularly sink nor rise, but like the smaller berries, be randomly distributed

The answer is b:

As the berries are shaken, gaps open up between and beneath them -- some large and some small gaps -- but mostly small gaps. Since only small berries can move down into the small gaps, over time the large berries are nudged to the top.

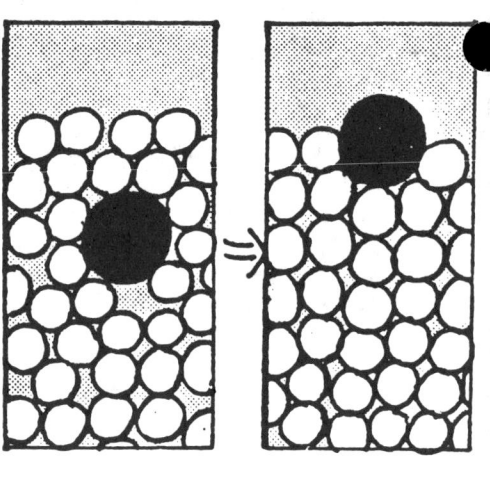

Interestingly enough, even denser objects move to the top in this way. Gentle motions in the ground nudge heavy rocks to the surface -- a source of frustration to gardeners.

CONCEPTUAL Physics

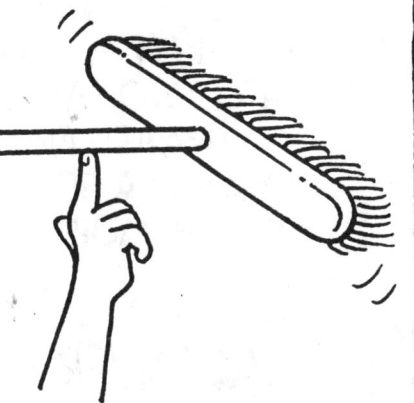

The broom balances at its center of gravity. If you saw the broom into two parts through the center of gravity and then weigh each part on a scale, which part will weigh more?

CONCEPTUAL Physics

THE BROOM BALANCES AT ITS CENTER OF GRAVITY. IF YOU SAW THE BROOM INTO TWO PARTS THROUGH THE CENTER OF GRAVITY AND THEN WEIGH EACH PART ON A SCALE, WHICH PART WILL WEIGH MORE?

ANSWER:

THE SHORT BROOM PART IS HEAVIER. IT BALANCES THE LONG HANDLE JUST AS KIDS OF UNEQUAL WEIGHTS CAN BALANCE ON A SEESAW WHEN THE HEAVIER KID SITS CLOSER TO THE FULCRUM. BOTH THE BALANCED BROOM AND SEESAW ARE EVIDENCE OF EQUAL AND OPPOSITE TORQUES --- NOT EQUAL WEIGHTS.

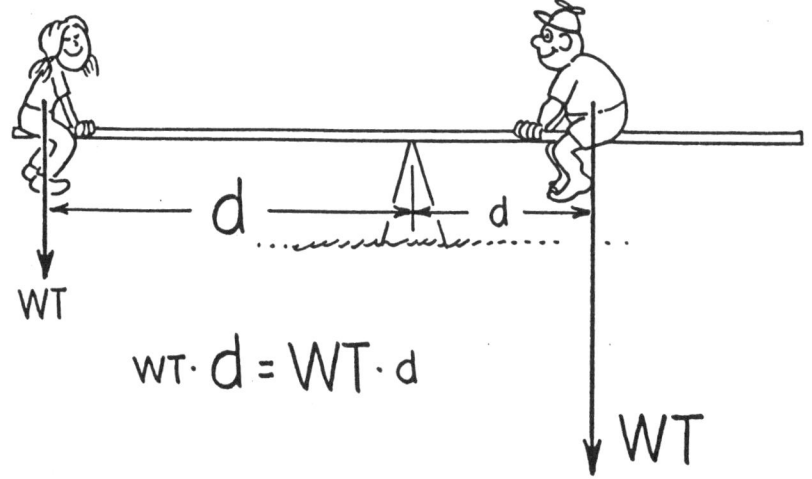

$$wt \cdot d = WT \cdot d$$

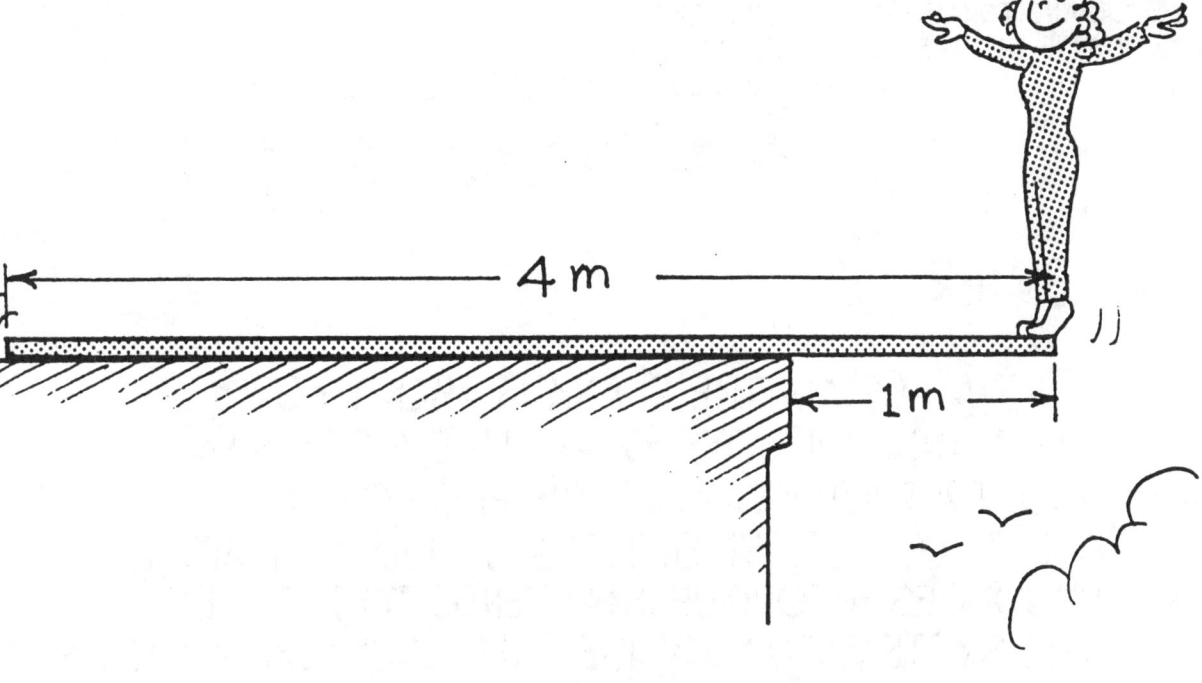

THE 40-kg WOMAN STANDS AT THE END OF A 4-METER-LONG UNIFORM PLANK. IF THE MAXIMUM OVERHANG FOR BALANCE IS 1 METER, ESTIMATE THE MASS OF THE PLANK.

CONCEPTUAL Physics

THE 40-kg WOMAN STANDS AT THE END OF A 4-METER-LONG UNIFORM PLANK. IF THE MAXIMUM OVERHANG FOR BALANCE IS 1 METER, ESTIMATE THE MASS OF THE PLANK.

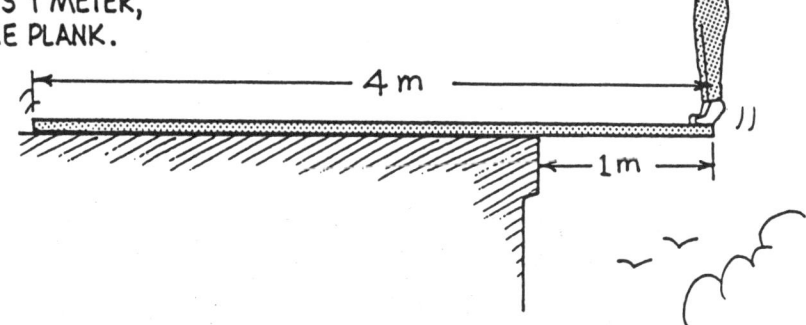

ANSWER:

THE MASS OF THE PLANK IS ABOUT 40 kg. THE PLANK TENDS TO ROTATE LIKE A SEE-SAW ABOUT A PIVOT POINT AT THE EDGE OF THE BUILDING. HER WEIGHT MULTIPLIED BY 1 METER PRODUCES A TORQUE THAT TENDS TO ROTATE THE SYSTEM CLOCKWISE. THE COUNTERBALANCING TORQUE IS PRODUCED BY THE WEIGHT OF THE PLANK MULTIPLIED BY THE DISTANCE FROM THE PIVOT POINT TO THE PLANK'S CENTER OF GRAVITY. NOTE THAT THIS DISTANCE IS ALSO 1 METER. SO BOTH THE WOMAN AND THE PLANK WEIGH THE SAME. THEIR MASSES ARE EQUAL.

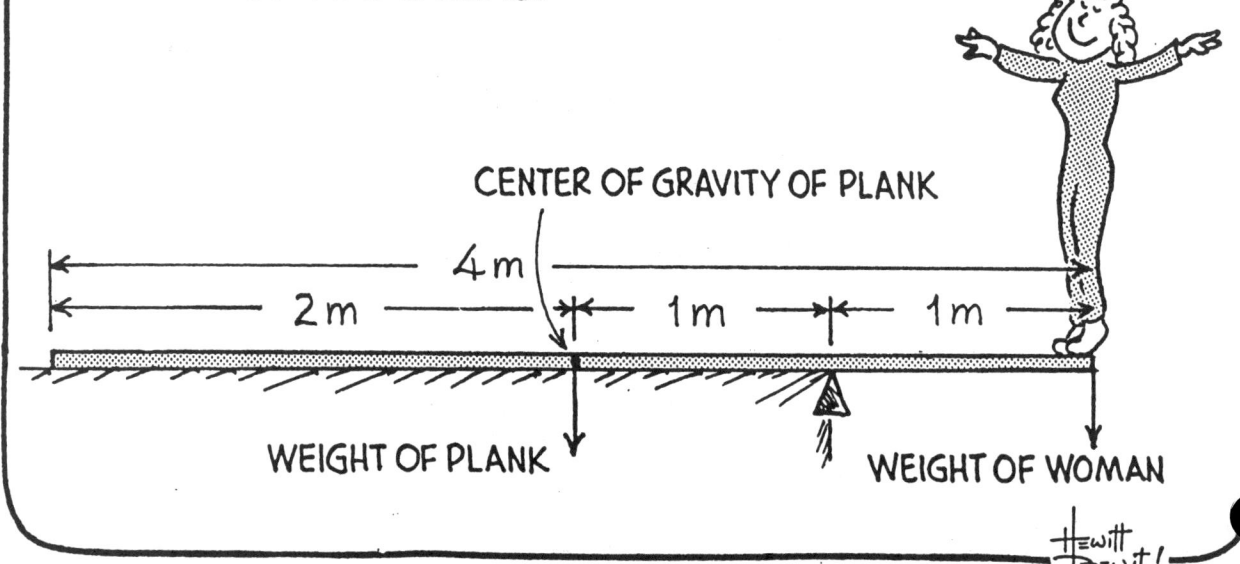

CONCEPTUAL Physics

With each revolution of the earth about its polar axis, we experience two high (and two low) ocean tides. If the moon were covered with water, would there similarly be two tides per lunar revolution?

EARTH

MOON

CONCEPTUAL Physics

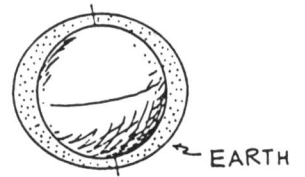
EARTH

WITH EACH REVOLUTION OF THE EARTH ABOUT ITS POLAR AXIS, WE EXPERIENCE TWO HIGH (AND TWO LOW) OCEAN TIDES. IF THE MOON WERE COVERED WITH WATER, WOULD THERE SIMILARLY BE TWO TIDES PER LUNAR REVOLUTION?

MOON

ANSWER:

NO. THE PRESENCE OF AN ASTRONOMICAL BODY NEAR ANOTHER PRODUCES A PAIR OF TIDAL BULGES. BUT WHETHER OR NOT THE BULGES RESULT IN "TIDES" (THE PERIODIC MOVING UP AND DOWN OF THE SURFACE) DEPENDS ON WHETHER OR NOT THE BODY ROTATES BENEATH THESE BULGES. SO IF THE MOON WERE COVERED WITH WATER THERE WOULD BE TWO TIDAL BULGES, LIKE ON EARTH AND FOR THE SAME REASON. BUT THERE WOULD BE NO PERIODIC HIGH AND LOW TIDES, BECAUSE THE MOON DOES NOT ROTATE BENEATH THESE BULGES. THE MOON ROTATES ONCE EACH 28 DAYS WITH RESPECT TO THE STARS, BUT DOES NOT ROTATE AT ALL WITH RESPECT TO THE EARTH -- SO THE BULGES, LIKE THE SIDE OF THE MOON THAT ALWAYS FACES THE EARTH, WOULD BE "FROZEN," WITH NO HIGH AND LOW TIDES TO SWEEP ACROSS THE MOON'S SURFACE.

CONCEPTUAL Physics

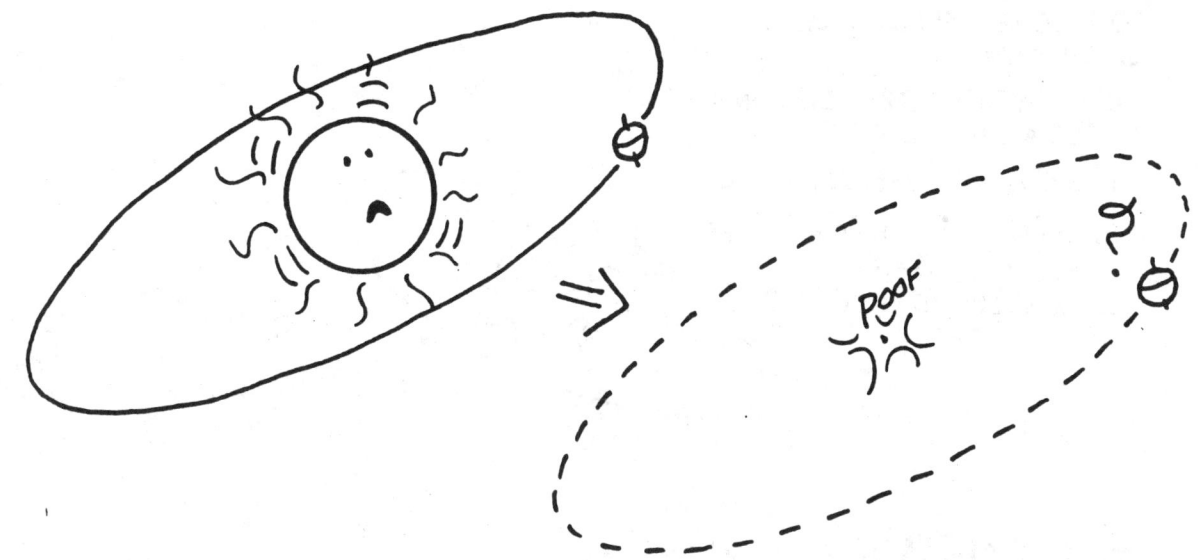

If the sun suddenly collapsed to become a black hole, the earth would

a) LEAVE THE SOLAR SYSTEM IN A STRAIGHT-LINE PATH

b) SPIRAL INTO THE BLACK HOLE

c) UNDERGO A MAJOR INCREASE IN TIDAL FORCES

d) CONTINUE TO CIRCLE IN ITS USUAL ORBIT

CONCEPTUAL Physics

IF THE SUN SUDDENLY COLLAPSED TO BECOME A BLACK HOLE, THE EARTH WOULD

a) LEAVE THE SOLAR SYSTEM IN A STRAIGHT-LINE PATH
b) SPIRAL INTO THE BLACK HOLE
c) UNDERGO A MAJOR INCREASE IN TIDAL FORCES
d) CONTINUE TO CIRCLE IN ITS USUAL ORBIT

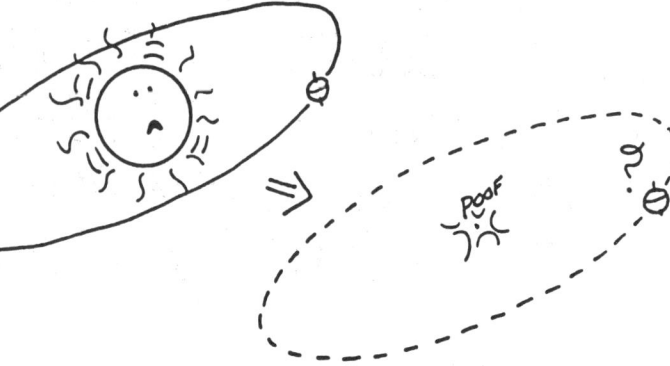

THE ANSWER IS d:

WE CAN SEE FROM NEWTON'S EQUATION,

$$F = G\frac{mM}{d^2}$$

THAT THE INTERACTION F BETWEEN THE MASS OF THE EARTH AND THE SUN DOESN'T CHANGE. THIS IS BECAUSE THE MASS OF THE EARTH DOES NOT CHANGE, THE MASS OF THE SUN DOES NOT CHANGE EVEN THOUGH IT IS COMPRESSED, AND THE DISTANCE FROM THE CENTERS OF THE EARTH AND THE SUN, COLLAPSED OR NOT, DOES NOT CHANGE. ALTHOUGH THE EARTH WOULD VERY SOON FREEZE AND UNDERGO ENORMOUS SURFACE CHANGES, ITS YEARLY PATH WOULD CONTINUE AS IF THE SUN WERE ITS NORMAL SIZE.

CONCEPTUAL Physics

Consider the various positions of the satellite as it orbits the planet as shown. With respect to the planet, in which position does the satellite have the maximum

a) SPEED?
b) VELOCITY?
c) MOMENTUM?
d) KINETIC ENERGY?
e) GRAVITATIONAL POTENTIAL ENERGY?
f) TOTAL ENERGY?

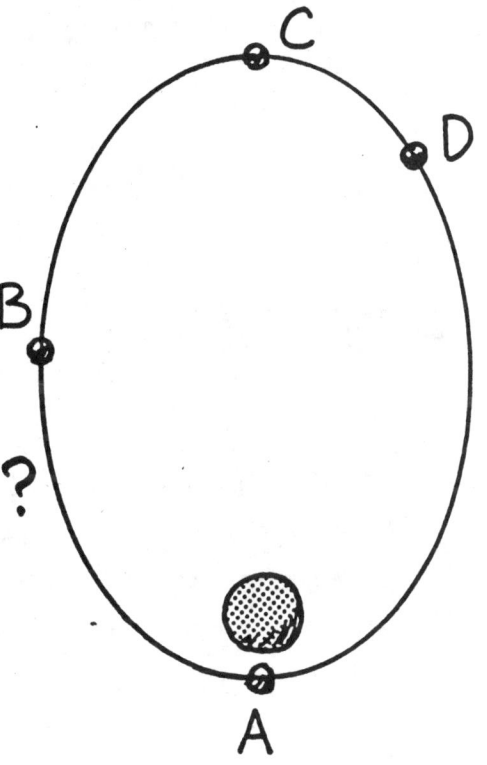

CONCEPTUAL Physics

Consider the various positions of the satellite as it orbits the planet as shown. With respect to the planet, in which position does the satellite have the maximum

a) SPEED?
b) VELOCITY?
c) MOMENTUM?
d) KINETIC ENERGY?
e) GRAVITATIONAL POTENTIAL ENERGY?
f) TOTAL ENERGY?

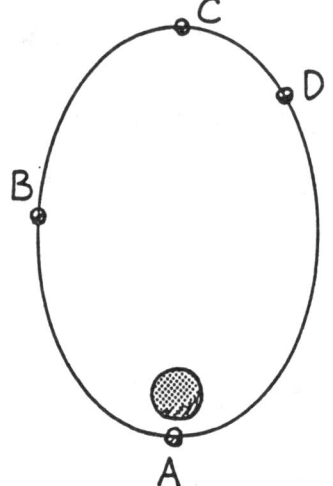

ANSWERS:

The satellite has greatest speed, velocity, momentum, and kinetic energy at the perigee, position **A**. It has greatest gravitational energy at the farthest position, the apogee at **C**. Total energy, KE + PE, is the same at all positions.

CONCEPTUAL Physics

A ROCKET FIRED VERTICALLY AT A SPEED OF 11.2 km/s WILL ESCAPE THE EARTH. IF IT IS INSTEAD LAUNCHED HORIZONTALLY AT THE SAME SPEED, AND IT DOESN'T HIT MOUNTAINS OR OTHER OBSTRUCTIONS, AND AIR RESISTANCE CAN BE NEGLECTED, WILL IT STILL ESCAPE THE EARTH?

CONCEPTUAL Physics

A ROCKET FIRED VERTICALLY AT A SPEED OF 11.2 km/s WILL ESCAPE THE EARTH. IF IT IS INSTEAD LAUNCHED HORIZONTALLY AT THE SAME SPEED, AND IT DOESN'T HIT MOUNTAINS OR OTHER OBSTRUCTIONS, AND AIR RESISTANCE CAN BE NEGLECTED, WILL IT STILL ESCAPE THE EARTH?

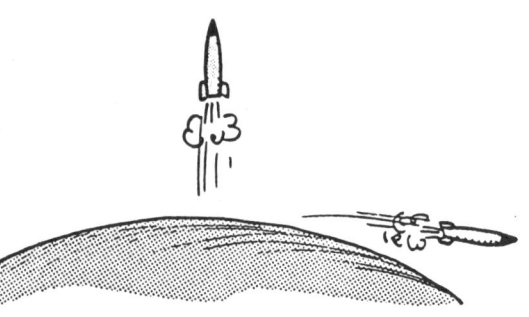

ANSWER:

YES. WHETHER OR NOT A BODY ESCAPES THE EARTH DEPENDS ON WHETHER OR NOT IT HAS SUFFICIENT KINETIC ENERGY TO EQUAL THE GRAVITATIONAL POTENTIAL ENERGY IT WOULD HAVE INFINITELY FAR AWAY. AT 11.2 km/s, THE ROCKET WILL HAVE THE SAME SUFFICIENT KINETIC ENERGY, WHETHER IT IS LAUNCHED VERTICALLY OR HORIZONTALLY.

CONCEPTUAL Physics

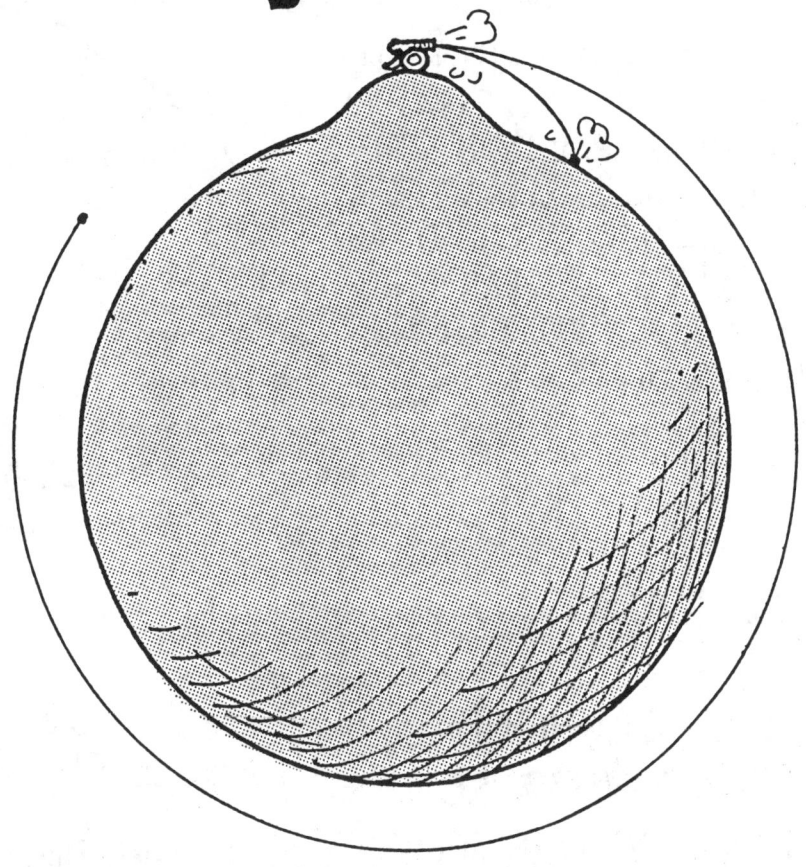

A CANNONBALL IS FIRED HORIZONTALLY FROM A TALL MOUNTAIN TO THE GROUND BELOW. BECAUSE OF GRAVITY, IT STRIKES THE GROUND WITH INCREASED SPEED. A SECOND CANNONBALL IS FIRED FAST ENOUGH TO GO INTO CIRCULAR ORBIT --- BUT GRAVITY DOES NOT INCREASE ITS SPEED. WHY?

CHAP. 9

CONCEPTUAL Physics

A CANNONBALL IS FIRED HORIZONTALLY FROM A TALL MOUNTAIN TO THE GROUND BELOW. BECAUSE OF GRAVITY, IT STRIKES THE GROUND WITH INCREASED SPEED. A SECOND CANNONBALL IS FIRED FAST ENOUGH TO GO INTO CIRCULAR ORBIT --- BUT GRAVITY DOES NOT INCREASE ITS SPEED. WHY?

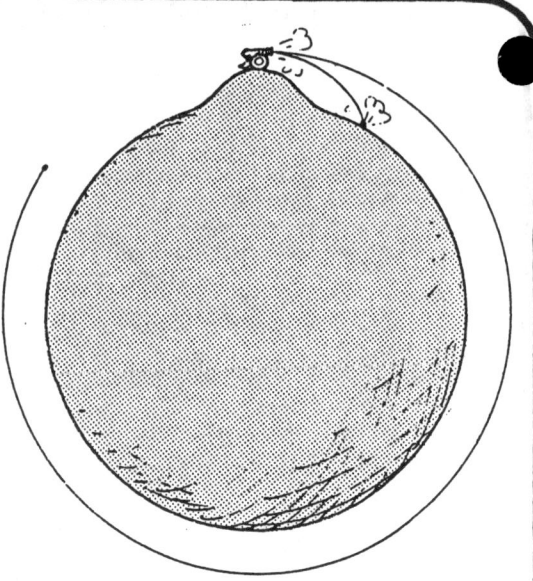

ANSWER:

THE FIRST CANNONBALL MOVES DOWNWARD, SO THERE IS A COMPONENT OF GRAVITATIONAL FORCE ALONG ITS DIRECTION OF MOTION THAT SPEEDS IT UP.

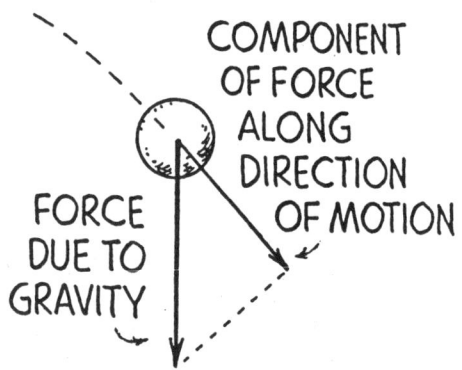

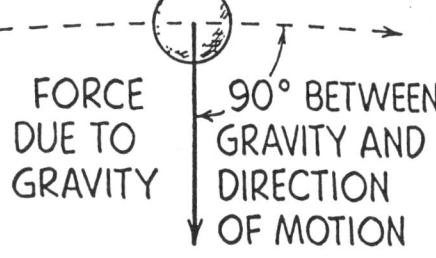

THE SECOND CANNONBALL MOVES PERPENDICULAR TO THE GRAVITATIONAL FORCE, WITH NO FORCE COMPONENT ALONG ITS DIRECTION OF MOTION. THAT'S WHY IT ORBITS AT CONSTANT SPEED.

CONCEPTUAL Physics

WOULD OBSERVERS ON THE MOON SEE THE EARTH "RISE" AND "SET", AS WE HERE ON EARTH SEE THE MOON RISE AND SET?

CHAP. 9

CONCEPTUAL Physics

Would observers on the moon see the earth "rise" and "set", as we here on earth see the moon rise and set?

ANSWER:

No. The rotation of the moon is synchronized with its revolution about the earth, with the result that the same side of the moon always faces the earth. So from the moon, the earth always appears at the same point in the sky. (Of course, the earth will be seen to go through the same phases, from crescent to full, as the moon does from earth.)

CHAP. 9

CONCEPTUAL Physics

Consider two satellites in orbit about a star (like our sun). If one satellite is twice as far from the star as the other, but both satellites are attracted to the star with the same gravitational force, how do the masses of the satellites compare?

CONCEPTUAL Physics

Consider two satellites in orbit about a star (like our sun). If one satellite is twice as far from the star as the other, but both satellites are attracted to the star with the same gravitational force, how do the masses of the satellites compare?

Answer:

If both satellites had the same mass, then the one twice as far would be attracted to the star with only one-fourth the force (inverse-square law). Since the force is the same for both, the mass of the farthermost satellite must be four times as great as the mass of the closer satellite.

$$\frac{mM}{d^2} = \frac{4mM}{(2d)^2}$$

CHAP. 9

CONCEPTUAL Physics

Which would require the greater change in the Earth's orbital speed (30 km/s); slowing it down so that it would crash into the sun, or speeding it up so that it would escape the sun?

Which would require the greater energy?

CONCEPTUAL Physics

WHICH WOULD REQUIRE THE GREATER CHANGE IN THE EARTH'S ORBITAL SPEED (30 km/s); SLOWING IT DOWN SO THAT IT WOULD CRASH INTO THE SUN, OR SPEEDING IT UP SO THAT IT WOULD ESCAPE THE SUN?

WHICH WOULD REQUIRE THE GREATER ENERGY?

ANSWERS:

IN ORDER TO CRASH INTO THE SUN, THE EARTH'S ORBITAL SPEED OF 30 km/s WOULD HAVE TO BE REDUCED TO ZERO. THIS IS A CHANGE OF 30 km/s. IN ORDER TO ESCAPE THE SUN, THE EARTH'S ORBITAL SPEED WOULD HAVE TO BE INCREASED TO 42.5 km/s, A CHANGE OF 12.5 km/s. SO A GREATER CHANGE IN SPEED IS REQUIRED TO SLOW THE EARTH FOR A SUN CRASH THAN TO SPEED IT FOR SOLAR ESCAPE.

BUT THE ENERGY NEEDED TO INCREASE IT FROM 30 km/s TO 42.5 km/s IS THE SAME AS THE ENERGY NEEDED TO SLOW IT TO ZERO. THAT IS,

$$\Delta KE_{(30 \rightarrow 42.5)} = \Delta KE_{(0 \rightarrow 30)}.$$

$$[½ M (42.5)^2 - ½ M (30)^2 \approx ½ M (30)^2 = ½ M (900)]$$

NOTE THAT $(42.5^2 - 30^2)$ IS NOT $(42.5 - 30)^2$!

(THE KE OF ANY SATELLITE IN CIRCULAR ORBIT IS HALF THE ENERGY NEEDED TO GET IT TO INFINITY!)

CONCEPTUAL Physics

A ROCKET COASTS IN AN ELLIPTICAL ORBIT AROUND THE EARTH.

TO ATTAIN ESCAPE VELOCITY USING THE LEAST AMOUNT OF FUEL IN A BRIEF FIRING TIME, SHOULD IT FIRE OFF AT THE APOGEE, OR AT THE PERIGEE?

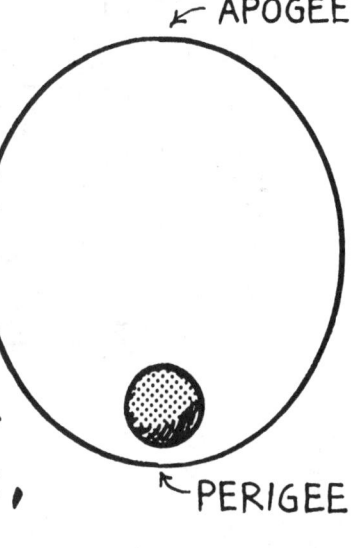

HINT: LET THE FORMULA

$$Fd = \Delta KE$$

GUIDE YOUR THINKING

CONCEPTUAL Physics

A ROCKET COASTS IN AN ELLIPTICAL ORBIT AROUND THE EARTH.

TO ATTAIN ESCAPE VELOCITY USING THE LEAST AMOUNT OF FUEL IN A BRIEF FIRING TIME, SHOULD IT FIRE OFF AT THE APOGEE, OR AT THE PERIGEE?

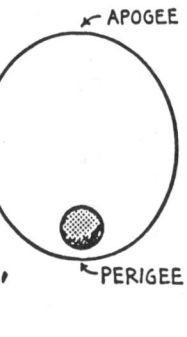

HINT: LET THE FORMULA
$$Fd = \Delta KE$$
GUIDE YOUR THINKING

ANSWER:

IN ACCORD WITH THE WORK-ENERGY FORMULA, $Fd = \Delta KE$, FOR A CONSTANT THRUST F, THE MAXIMUM CHANGE IN KE WILL OCCUR WHEN d IS MAXIMUM. THE ROCKET WILL TRAVEL THE GREATEST DISTANCE d DURING THE BRIEF FIRING TIME WHERE IT IS TRAVELING FASTEST -- AT THE *PERIGEE*.

THIS CAN BE SEEN ALSO BY CONSIDERING THE RELATIVE KEs GIVEN TO THE EXHAUST GASES AT THE PERIGEE AND APOGEE. AT THE APOGEE, WHERE THE ROCKET COASTS SLOWER, MUCH MORE KE OF THE SYSTEM GOES TO THE GASES, WHEREAS AT THE PERIGEE MOST OF THE KE IS ASSOCIATED WITH THE ROCKET. (IF ORBITAL SPEED = ROCKET EXHAUST SPEED, THE GASES ARE MOTIONLESS WITH RESPECT TO THE EARTH AND THE ROCKET GETS 100% OF THE KE.)

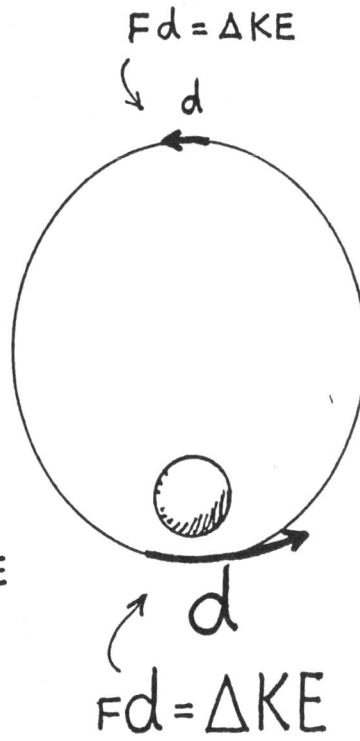

CONCEPTUAL Physics

Oops! Those "harmless" germanium capsules you just swallowed may have an extra proton in each nucleus.

Is this good news or bad news? Why?

CHAP. 10

CONCEPTUAL Physics

Oops! Those "harmless" germanium capsules you just swallowed may have an extra proton in each nucleus.

Is this good news or bad news? Why?

ANSWER:

This is bad news, for a germanian nucleus with an additional proton is not germanium, but arsenic!

Check the periodic table --- germanium is atomic number 32, and 33 is arsenic.

CONCEPTUAL Physics

> Suppose you could add or subtract protons from oxygen nuclei. To turn oxygen into a gas that would glow red when an electric current flows through it, would you add or subtract protons? How many?

ANSWER:

ADD TWO PROTONS TO EACH NUCLEUS OF OXYGEN AND YOU INCREASE THE ATOMIC NUMBER FROM 8 TO 10. YOU THEN HAVE NEON, WHICH WILL GLOW A VERY NICE RED WHEN A CURRENT FLOWS THROUGH IT.

CHAP 10

CONCEPTUAL **Physics**

Consider an infant who weighs 100 newtons. During a year she grows so that each dimension of her body increases by 5%.

How much will she then weigh? (Assume her density remains unchanged.)

CONCEPTUAL Physics

CONSIDER AN INFANT WHO WEIGHS 100 NEWTONS. DURING A YEAR SHE GROWS SO THAT EACH DIMENSION OF HER BODY INCREASES BY 5%.

HOW MUCH WILL SHE THEN WEIGH? (ASSUME HER DENSITY REMAINS UNCHANGED.)

ANSWER:

HER WEIGHT INCREASES BY 16% AND SHE WEIGHS 116 NEWTONS. THIS IS BECAUSE A 5% INCREASE MEANS THAT EACH DIMENSION OF HER BODY INCREASES TO 1.05 WHAT IT WAS THE YEAR BEFORE. SO THE SCALING FACTOR IS 1.05. HER WEIGHT INCREASES IN PROPORTION TO THE CUBE OF THIS SCALING FACTOR:

$$1.05 \times 1.05 \times 1.05 = 1.16$$

SO SHE IS 1.16 TIMES HEAVIER THAN THE YEAR BEFORE.

CONCEPTUAL Physics

Compared to an empty ship, will a ship loaded with a cargo of styrofoam float lower in water or higher in water?

ANSWER:

The ship loaded with styrofoam will float lower in water. A ship will float highest when its weight is least --- that is, when it is empty. Loading any cargo will increase its weight and make it float lower in the water. Whether the cargo is a ton of styrofoam or a ton of iron, the water displacement will be the same.

CONCEPTUAL Physics

EVERYBODY KNOWS THAT "WATER SEEKS ITS OWN LEVEL," BUT VERY FEW PEOPLE KNOW **WHY** WATER SEEKS ITS OWN LEVEL. THE REASON HAS MOST TO DO WITH

a) ATMOSPHERIC PRESSURE
b) WATER PRESSURE DEPENDING ON DEPTH
c) WATER'S DENSITY

CONCEPTUAL Physics

EVERYBODY KNOWS THAT "WATER SEEKS ITS OWN LEVEL." BUT VERY FEW PEOPLE KNOW **WHY** WATER SEEKS ITS OWN LEVEL. THE REASON HAS MOST TO DO WITH

a) ATMOSPHERIC PRESSURE
b) WATER PRESSURE DEPENDING ON DEPTH
c) WATER'S DENSITY

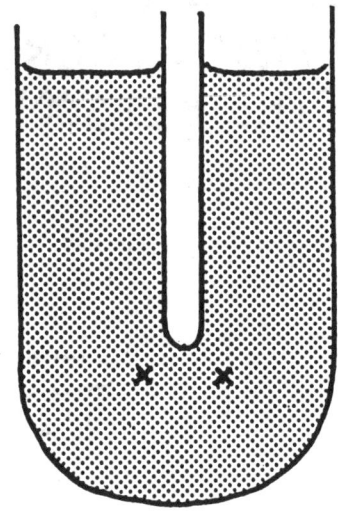

THE ANSWER IS b:

WATER PRESSURE DEPENDS ON DEPTH, SO ONLY AT EQUAL DEPTHS OF WATER WILL THE PRESSURE BE EQUAL. CONSIDER THE U-TUBE. IF WATER IS AT REST WHERE EACH X IS, THE PRESSURES MUST BE EQUAL -- OTHERWISE A FLOW WOULD OCCUR FROM THE REGION OF HIGHER TO THE REGION OF LOWER PRESSURE UNTIL THE PRESSURES EQUALIZE. FOR THIS TO HAPPEN, THE DEPTHS BELOW THE SURFACES MUST BE EQUAL.

THIS IS TRUE WHATEVER THE DENSITY OF WATER OR WHETHER OR NOT THERE IS ATMOSPHERIC PRESSURE.

CHAP. 12

CONCEPTUAL Physics

Consider a boat loaded with scrap iron in a swimming pool. If the iron is thrown overboard into the pool, will the water level at the edge of the pool rise, fall, or remain unchanged?

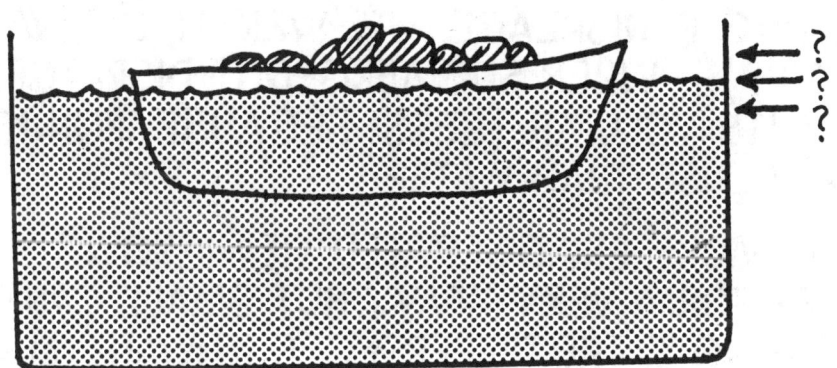

CHAP. 12

CONCEPTUAL Physics

Consider a boat loaded with scrap iron in a swimming pool. If the iron is thrown overboard into the pool, will the water level at the edge of the pool rise, fall, or remain unchanged?

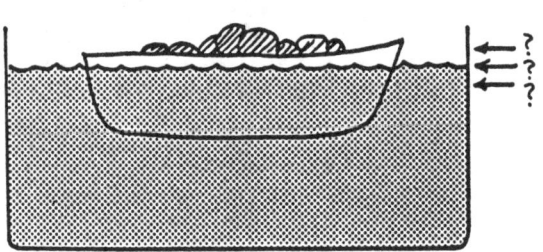

ANSWER:

The water level at the side of the pool will fall, because the iron will displace less water submerged than when floating. When floating it displaces its weight of water (a lot!) --- when submerged it displaces only its volume (less, because iron is more dense than water).

The more exaggerated view shows cases for a very heavy but small cannonball --- note the differences in water levels.

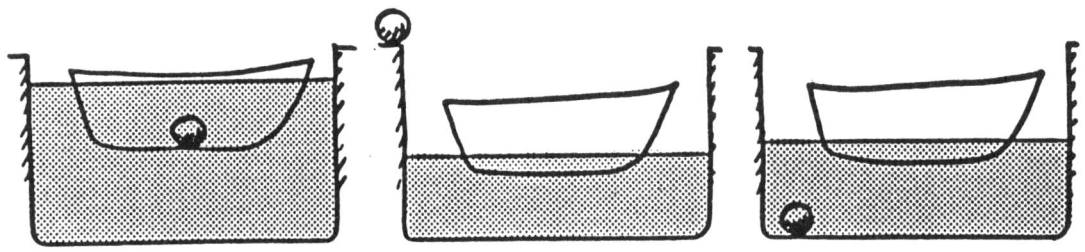

CHAP. 12

CONCEPTUAL Physics

A BLOCK OF BALSA WOOD WITH A ROCK TIED TO IT FLOATS IN WATER. WHEN THE ROCK IS ON TOP AS SHOWN, EXACTLY HALF THE BLOCK IS BELOW THE WATER LINE. WHEN THE BLOCK IS TURNED OVER SO THE ROCK IS UNDERNEATH AND SUBMERGED, THE AMOUNT OF BLOCK BELOW THE WATER LINE IS

a) LESS THAN HALF
b) HALF
c) MORE THAN HALF

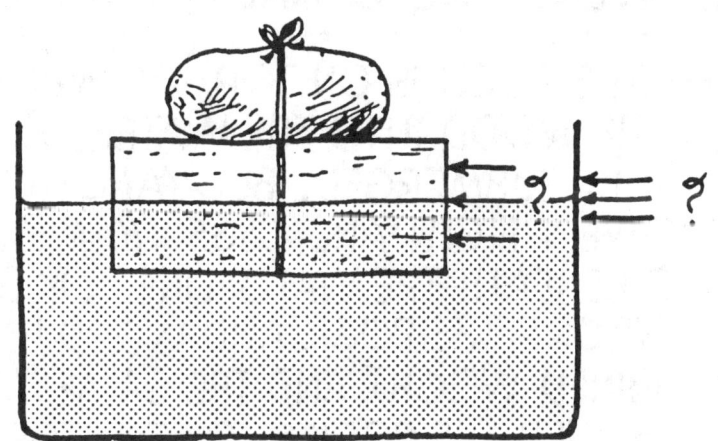

AND THE WATER LEVEL AT THE SIDE OF THE CONTAINER WILL

d) RISE
e) FALL
f) REMAIN UNCHANGED

CONCEPTUAL Physics

A BLOCK OF BALSA WOOD WITH A ROCK TIED TO IT FLOATS IN WATER. WHEN THE ROCK IS ON TOP AS SHOWN, EXACTLY HALF THE BLOCK IS BELOW THE WATER LINE. WHEN THE BLOCK IS TURNED OVER SO THE ROCK IS UNDERNEATH AND SUBMERGED, THE AMOUNT OF BLOCK BELOW THE WATER LINE IS

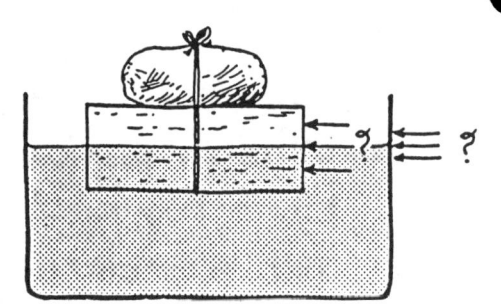

a) LESS THAN HALF
b) HALF
c) MORE THAN HALF

AND THE WATER LEVEL AT THE SIDE OF THE CONTAINER WILL

d) RISE
e) FALL
f) REMAIN UNCHANGED

THE ANSWERS ARE a AND f:

WHEN THE ROCK IS ON TOP, ITS WHOLE WEIGHT PUSHES THE WOOD INTO THE WATER. BUT WHEN THE ROCK IS SUBMERGED, BUOYANCY ON IT REDUCES ITS EFFECTIVE WEIGHT AND *LESS THAN HALF THE BLOCK* IS PULLED BENEATH THE WATER LINE. OR BY THE LAW OF FLOTATION: THE ROCK AND WOOD UNIT DISPLACES ITS COMBINED WEIGHT AND THE SAME VOLUME OF WATER WHETHER THE ROCK IS ON THE TOP OR THE BOTTOM. WHEN THE ROCK IS ON THE BOTTOM, LESS WOOD IS BELOW THE WATER LINE THAN WHEN THE ROCK IS ON THE TOP.

SINCE THE SAME VOLUME OF WATER IS DISPLACED NO MATTER HOW IT FLOATS, THE WATER LEVEL AT THE SIDE OF THE CONTAINER REMAINS *UNCHANGED*.

CONCEPTUAL Physics

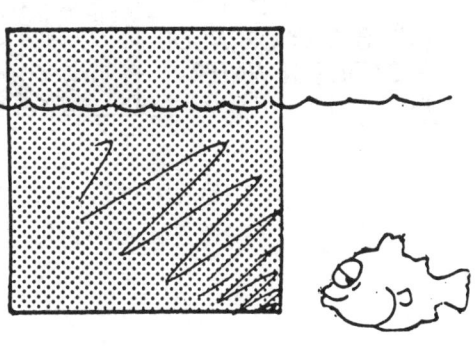

THE DENSITY OF THE BLOCK OF WOOD FLOATING IN WATER IS

a) GREATER THAN THE DENSITY OF WATER

b) EQUAL TO THE DENSITY OF WATER

c) LESS THAN HALF THAT OF WATER

d) MORE THAN HALF THE DENSITY OF WATER

e) ... NOT ENOUGH INFORMATION IS GIVEN

CONCEPTUAL Physics

THE DENSITY OF THE BLOCK OF WOOD FLOATING IN WATER IS

a) GREATER THAN THE DENSITY OF WATER
b) EQUAL TO THE DENSITY OF WATER
c) LESS THAN HALF THAT OF WATER
d) MORE THAN HALF THE DENSITY OF WATER
e) ... NOT ENOUGH INFORMATION IS GIVEN

THE ANSWER IS d:

A VERY-LOW DENSITY OBJECT, LIKE AN INFLATED BALLOON, FLOATS HIGH ON THE WATER, AND A DENSER OBJECT LIKE A PIECE OF HARDWOOD, FLOATS LOWER INTO THE WATER. AN OBJECT HALF AS DENSE AS WATER FLOATS HALF WAY INTO THE WATER (BECAUSE IT WEIGHS AS MUCH AS HALF ITS VOLUME OF WATER). WOOD THAT FLOATS 3/4 SUBMERGED, IS 3/4 AS DENSE AS WATER --- LIKE THE BLOCK IN QUESTION --- MORE THAN HALF THE DENSITY OF WATER.

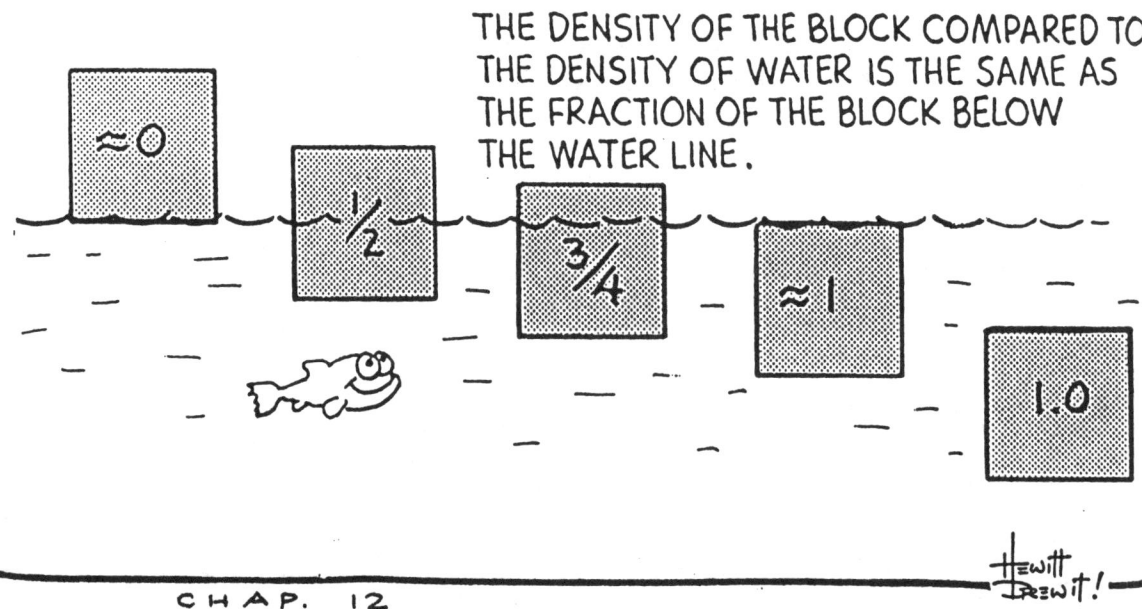

THE DENSITY OF THE BLOCK COMPARED TO THE DENSITY OF WATER IS THE SAME AS THE FRACTION OF THE BLOCK BELOW THE WATER LINE.

CONCEPTUAL Physics

The weight of the stand and suspended solid iron ball is equal to the weight of the container of water as shown above. When the ball is lowered into the water, the balance is upset. The amount of weight that must be added to the left side to restore balance, compared to the weight of water displaced by the ball, would be

a) half

b) the same

c) twice

d) more than twice

CHAP. 12

CONCEPTUAL Physics

The weight of the stand and suspended solid iron ball is equal to the weight of the container of water as shown above. When the ball is lowered into the water, the balance is upset. The amount of weight that must be added to the left side to restore balance, compared to the weight of water displaced by the ball, would be

a) half
b) the same
c) twice
d) more than twice

The answer is C:

The additional weight that must be put on the left side to restore balance will equal twice the buoyant force, that is, twice the weight of water displaced by the submerged ball. Why twice? Because what the right side gains because of submersion and heightened water level, the left side loses. (For example, if each side initially weighs 10 N and the right side gains 2 N to become 12 N, the left side loses 2 N to become 8 N. So an additional weight of 4 N, not 2 N, is required on the left side to restore balance.)

CHAP. 12

CONCEPTUAL **Physics**

WHY IS WET SAND FIRMER BENEATH YOUR FEET THAN WET GRAVEL?

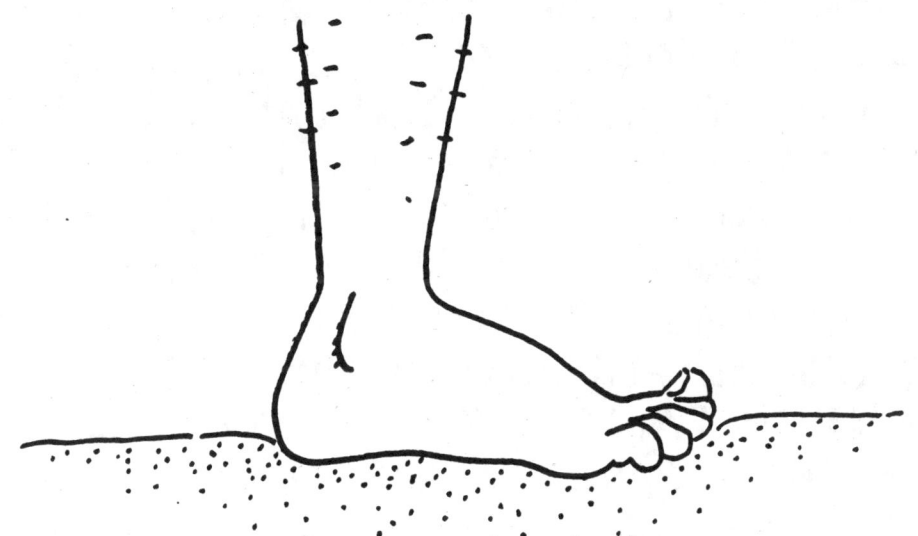

CHAP 12

CONCEPTUAL Physics

WHY IS WET SAND FIRMER BENEATH YOUR FEET THAN WET GRAVEL?

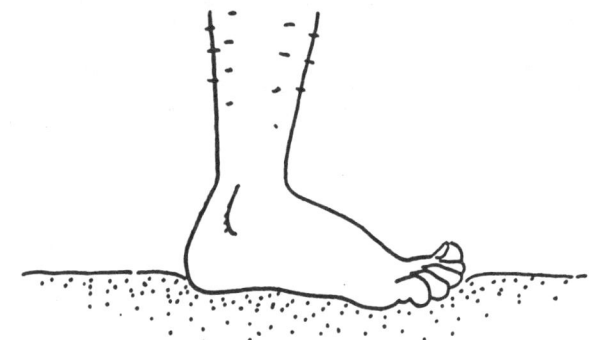

ANSWER:

THERE IS CONSIDERABLY MORE SURFACE AREA PER KILOGRAM OF SAND THAN PER KILOGRAM OF GRAVEL. YOU CAN SEE THIS BY COMPARING THE AMOUNT OF PAINT IT WOULD TAKE TO PAINT A 1-KILOGRAM ROCK WITH THE LARGER AMOUNT OF PAINT IT WOULD TAKE TO COVER ITS PIECES IF IT WERE BROKEN INTO GRAVEL; THEN WITH THE STILL GREATER AMOUNT OF PAINT THAT WOULD BE SOAKED UP IF THE GRAVEL WERE GROUND INTO SAND. THE GREATER SURFACE AREA WHEN WETTED WITH WATER PRODUCES MORE SURFACE TENSION THAT PULLS THE GRAINS TOGETHER TO FORM FIRM SAND.

CONCEPTUAL Physics

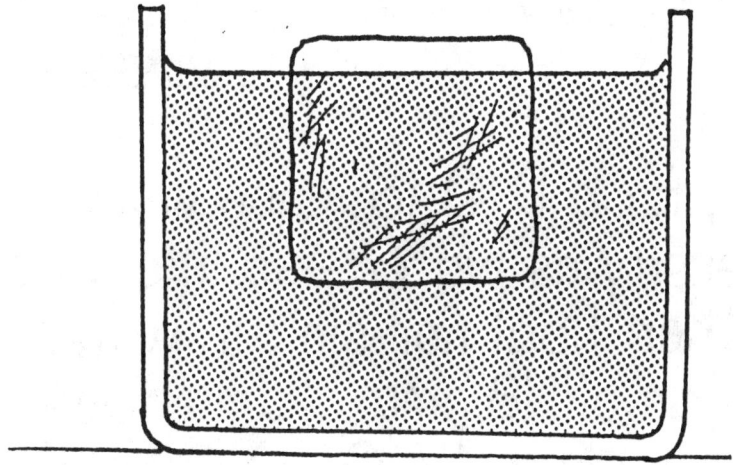

An ice cube of solid H_2O floats in water. Suppose all the hydrogen in the cube were the heavy isotope of hydrogen, deuterium (H-2) — would the ice cube of solid D_2O float or sink in ordinary water?

CONCEPTUAL Physics

An ice cube of solid H_2O floats in water. Suppose all the hydrogen in the cube were the heavy isotope of hydrogen, deuterium (H-2) — would the ice cube of solid D_2O float or sink in ordinary water?

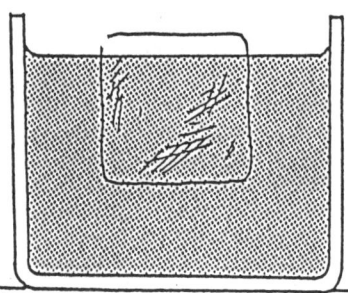

Answer:

Neither, for a D_2O ice cube will have the same density as water — so like a fish, it will neither float nor sink in water. H_2O has a mass of 18 amu, 1 amu for each H and 16 amu for the O. D_2O on the other hand has a mass of 20 amu, because of 2 amu for each D. The D_2O cube is $20/18 = 10/9$ times heavier, and therefore $10/9$ times denser (because only the weight and not the volumes of the cubes is different). The density of ordinary ice is $9/10$ that of water. So we see that $10/9$ of $9/10 = 1$, the same density of water.

> Both ordinary water and heavy water expand the same when they become ice. Since the increased volume of ice is exactly matched by the greater mass of heavy water, heavy-water ice has the same density as ordinary water.

CHAP. 12

CONCEPTUAL Physics

Consider a flexible plastic bottle containing both air and water immersed neck down in an open dish of water. The water level in the bottle will

a) fall if pinched at A but rise if pinched at B

b) fall if pinched at A or at B

c) fall if pinched at A but stay where it is if pinched at B

d) rise if pinched at A but stay where it is if pinched at B

e) stay where it is if pinched at A or at B

CONCEPTUAL Physics

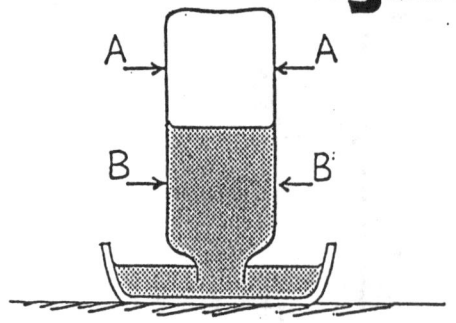

Consider a flexible plastic bottle containing both air and water immersed neck down in an open dish of water. The water level in the bottle will

a) fall if pinched at A but rise if pinched at B
b) fall if pinched at A or at B
c) fall if pinched at A but stay where it is if pinched at B
d) rise if pinched at A but stay where it is if pinched at B
e) stay where it is if pinched at A or at B

The answer is C:

Pinching the bottle at A compresses the air within the bottle, which pushes water out the neck of the bottle into the open dish until air pressure inside and outside the bottle is practically the same. The water level in the bottle is lowered. Pinching the bottle at B simply forces water out the neck and into the open dish, rather than rising and compressing the air above. Again, air pressure inside and outside the bottle is the same.

After thinking about this, did you experiment? If so, place a gold star on your forehead!

CONCEPTUAL Physics

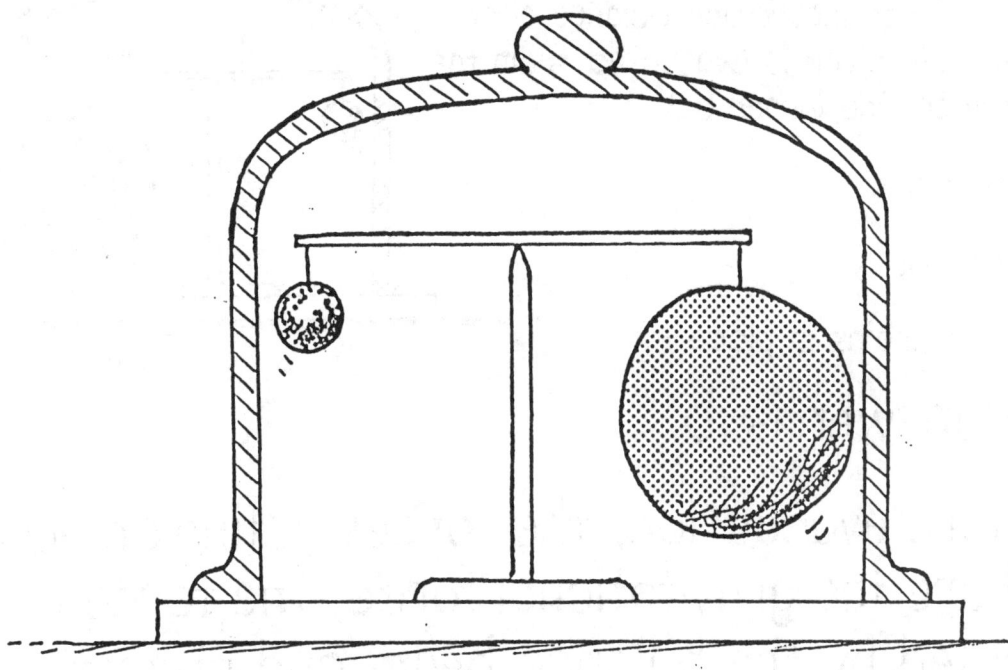

In the presence of air, the small iron ball and large plastic ball balance each other. When air is evacuated from the container, the larger ball

    a) rises

    b) falls

    c) remains in place

CONCEPTUAL Physics

In the presence of air, the small iron ball and large plastic ball balance each other. When air is evacuated from the container, the larger ball

a) rises

b) falls

c) remains in place

The answer is **b**:

Before evacuation, the forces acting on each ball are the gravitational force, the force exerted by the balance beam and the upward buoyant force exerted by the surrounding air. Evacuating the container removes the buoyant force on each ball. Since buoyant force equals the weight of air displaced, and the larger ball displaces the greater weight of air, the loss of buoyant force is greater for the larger ball, which falls.

> In the presence of air, a larger object with its greater buoyant force must have a slightly greater weight to balance a smaller object with its correspondingly-smaller buoyant force.

CHAP. 13

CONCEPTUAL Physics

Consider an air-filled balloon weighted so that it is on the verge of sinking --- that is, its overall density just equals that of water.

Now if you push it beneath the surface, it will

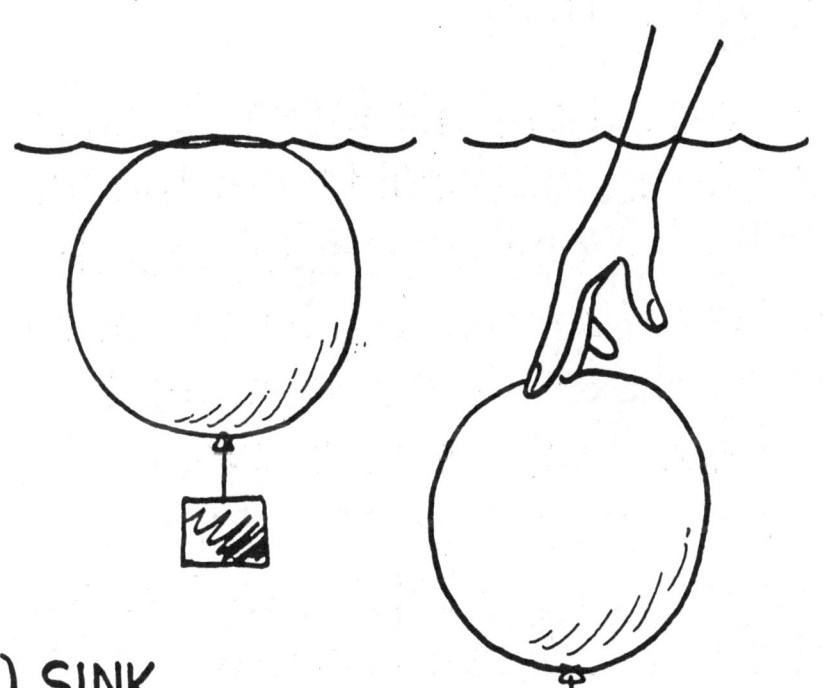

a) SINK

b) RETURN TO THE SURFACE

c) STAY AT THE DEPTH TO WHICH IT IS PUSHED

CONCEPTUAL Physics

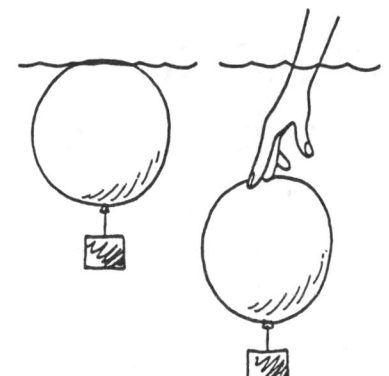

Consider an air-filled balloon weighted so that it is on the verge of sinking --- that is, its overall density just equals that of water.

Now if you push it beneath the surface, it will

a) SINK
b) RETURN TO THE SURFACE
c) STAY AT THE DEPTH TO WHICH IT IS PUSHED

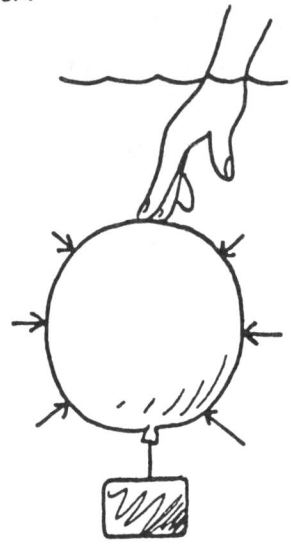

THE ANSWER IS a:

The balloon will sink. Why? Because at deeper levels the surrounding water pressure is greater and will squeeze and compress the balloon --- its density increases. Greater density results in sinking.

Or look at it this way: At the surface its buoyant force is just adequate for equilibrium. When the balloon is compressed it displaces less water and the buoyant force is reduced --- inadequate for equilibrium.

QUESTION: As the weighted balloon sinks, will the buoyant force increase, decrease, or remain the same?

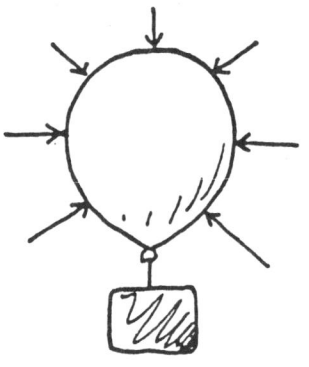

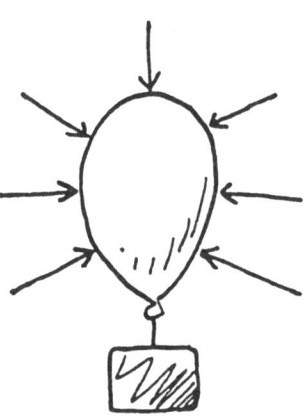

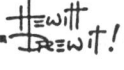

CHAP. 13

CONCEPTUAL Physics

If you release a ball inside a freely-falling elevator, it stays in front of you instead of "falling to the floor" because you, the ball, and the elevator are all accelerating downward at the same acceleration, **g**. If you similarly release a helium-filled balloon, the balloon will

a) also stay in front of you

b) press against the ceiling

c) press against the floor

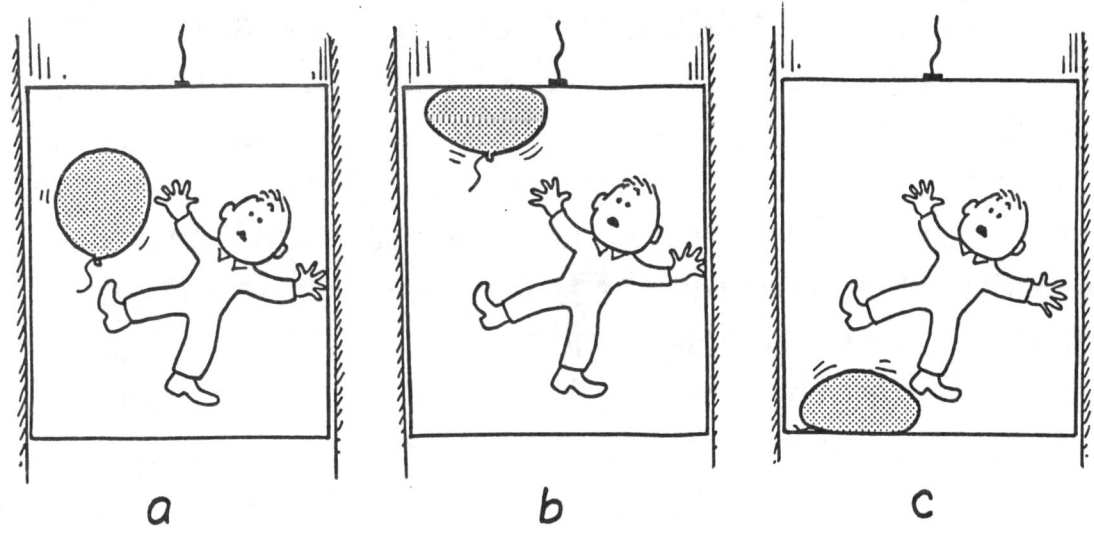

CHAP. 13

CONCEPTUAL Physics

If you release a ball inside a freely-falling elevator, it stays in front of you instead of "falling to the floor" because you, the ball, and the elevator are all accelerating downward at the same acceleration, g. If you similarly release a helium-filled balloon, the balloon will

a) also stay in front of you
b) press against the ceiling
c) press against the floor

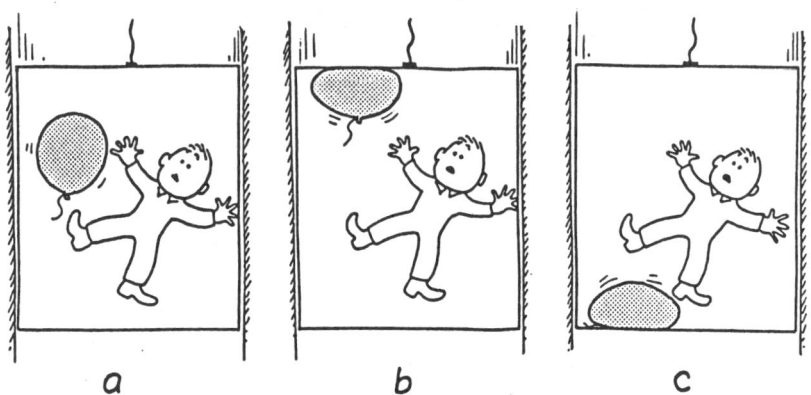

The answer is a:

Like the falling ball, the balloon will also stay in front of you because it loses its buoyancy. The buoyancy of a balloon is the result of a difference in air pressure against it -- usually a greater pressure acting up against the bottom than down against the top. But in the freely-falling environment there is no difference in air pressure. Air in the elevator like everything else inside, is in a state of free fall. Air in the top part does not press against air in the bottom part to give a greater pressure there. No pressure difference means no buoyancy, so the balloon freely falls like everything else inside the elevator.

CONCEPTUAL Physics

Compared to the mass of a dozen eggs, the mass of air in an "empty refrigerator" is

a) negligible
b) about a tenth as much
c) about the same
d) more

CONCEPTUAL Physics

Compared to the mass of a dozen eggs, the mass of air in an "empty refrigerator" is

a) negligible
b) about a tenth as much
c) about the same
d) more

Answer:

One cubic meter of air at 0°C and normal atmospheric pressure has a mass of about 1.3 kilograms. A medium-sized refrigerator has a volume of about 0.6 cubic meter and contains about 0.8 kilograms of air — more than the 0.75 kilograms of a dozen large eggs!

We don't notice the weight of air because we are submerged in air. If someone handed you a bag of water while you were submerged in water, you wouldn't notice its weight either.

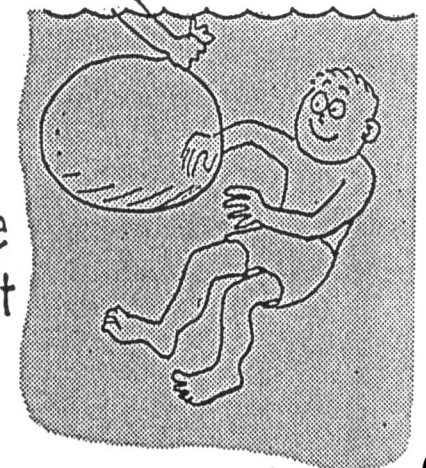

CONCEPTUAL Physics

A block of wood and a block of iron on weighing scales each weigh 1 ton. Strictly speaking, which has the greater mass?

Take buoyancy of air into account!

CHAP. 13

CONCEPTUAL Physics

A block of wood and a block of iron on weighing scales each weigh 1 ton. Strictly speaking, which has the greater mass?

Answer:

The wood has the greater mass. Why? Because the scale reading is weight *minus* the buoyant force of the surrounding air. The wood has a greater volume, displaces more air, and therefore has a greater buoyant force. To yield the same scale reading it must therefore have a greater mass than the iron.

How much greater?

By an amount that equals the difference in buoyant forces on the wood and iron blocks.

CONCEPTUAL Physics

The inverted drinking glass filled with air is placed mouth downward in water. As it is pushed deeper, the air is compressed. How deep must the glass be pushed in order that the air be compressed to half its original volume?

At this depth, how will the buoyant force on the submerged glass compare to when it was submerged at the surface?

CHAP. 13

CONCEPTUAL Physics

THE INVERTED DRINKING GLASS FILLED WITH AIR IS PLACED MOUTH DOWNWARD IN WATER. AS IT IS PUSHED DEEPER, THE AIR IS COMPRESSED. HOW DEEP MUST THE GLASS BE PUSHED IN ORDER THAT THE AIR BE COMPRESSED TO HALF ITS ORIGINAL VOLUME?

AT THIS DEPTH, HOW WILL THE BUOYANT FORCE ON THE SUBMERGED GLASS COMPARE TO WHEN IT WAS SUBMERGED AT THE SURFACE?

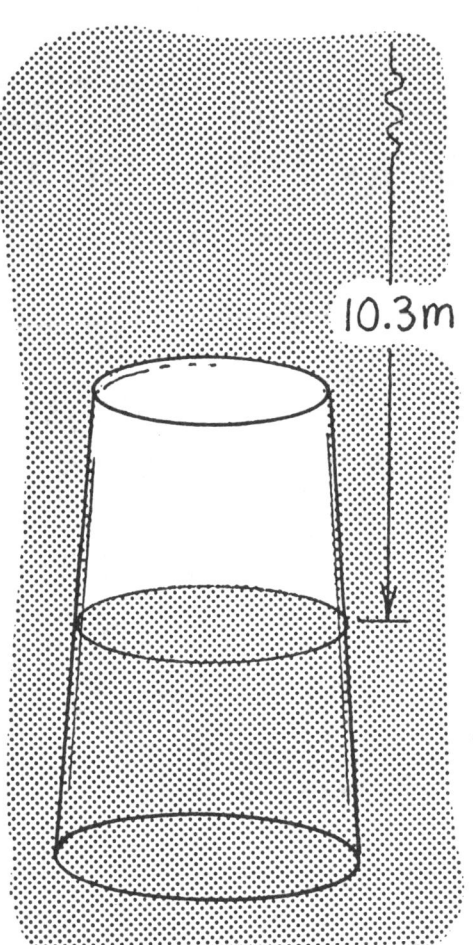

ANSWER:

THE AIR IN THE GLASS WILL BE SQUEEZED TO HALF VOLUME WHEN IT IS PUSHED 10.3 METERS BENEATH THE SURFACE. AT THIS DEPTH THE PRESSURE DUE TO WATER IS EQUAL TO THE PRESSURE OF THE ATMOSPHERE AT THE SURFACE. THIS MEANS THE PRESSURE ON THE AIR IS TWICE AT THIS DEPTH. TWICE THE PRESSURE, THEN HALF THE VOLUME.

HALF THE VOLUME MEANS HALF AS MUCH WATER IS DISPLACED BY THE GLASS, SO THE BUOYANT FORCE ON IT IS <u>HALF</u> THAT NEAR THE SURFACE.

CONCEPTUAL Physics

A BIRTHDAY CANDLE BURNS IN A DEEP DRINKING GLASS. WHEN THE GLASS IS WHIRLED AROUND IN A CIRCULAR PATH, SAY HELD AT ARM'S LENGTH WHILE ONE IS SPINNING LIKE AN ICE SKATER, WHICH WAY DOES THE CANDLE FLAME POINT?

CHAP. 13

CONCEPTUAL Physics

A BIRTHDAY CANDLE BURNS IN A DEEP DRINKING GLASS. WHEN THE GLASS IS WHIRLED AROUND IN A CIRCULAR PATH, SAY HELD AT ARM'S LENGTH WHILE ONE IS SPINNING LIKE AN ICE SKATER, WHICH WAY DOES THE CANDLE FLAME POINT?

ANSWER:

THE CANDLE FLAME POINTS INWARD, TOWARD THE CENTER OF THE CIRCULAR MOTION. THIS IS BECAUSE THE AIR IN THE GLASS IS MORE DENSE THAN THE FLAME AND "SLOSHES" TO THE FARTHER PART OF THE GLASS. THE GREATER AIR PRESSURE AT THE FARTHER PART OF THE INNER GLASS THEN BUOYS THE FLAME TO THE REGION OF LESSER PRESSURE--- INWARD.

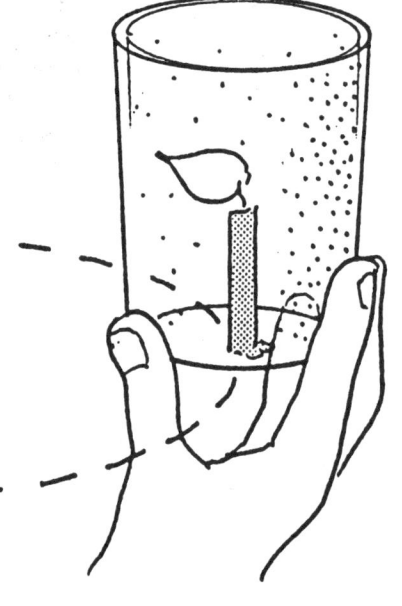

CHAP. 13

CONCEPTUAL Physics

A CANDLE WILL STAY LIT INSIDE THE SPACE SHUTTLE WHEN IT IS ON THE LAUNCH PAD, BUT NOT WHEN IT IS IN ORBIT. WHY?

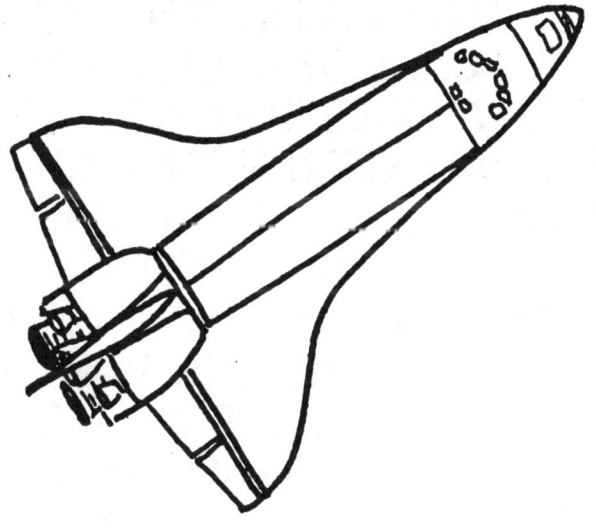

CHAP. 13

CONCEPTUAL Physics

A CANDLE WILL STAY LIT INSIDE THE SPACE SHUTTLE WHEN IT IS ON THE LAUNCH PAD, BUT NOT WHEN IT IS IN ORBIT. WHY?

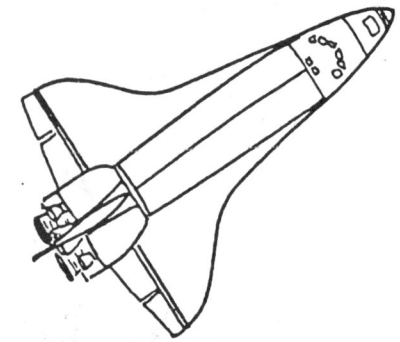

ANSWER:

WHEN A CANDLE ORDINARILY BURNS, THE WARMED CARBON DIOXIDE PRODUCED IN THE FLAME RISES BY CONVECTION, AND OXYGEN COMES IN FROM BELOW TO KEEP THE PROCESS GOING. BUT WHEN IN ORBIT, THERE IS NO EFFECT OF GRAVITY INSIDE THE CABIN AND CONVECTION CANNOT OCCUR. THE WARMED EXHAUST GASES DO NOT "RISE", AND INSTEAD SUFFOCATE THE FLAME.

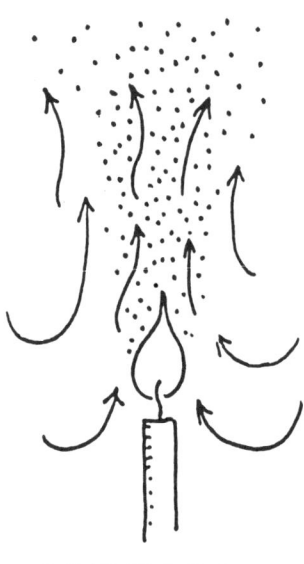

CONVECTION

NO CONVECTION

CHAP 13

CONCEPTUAL Physics

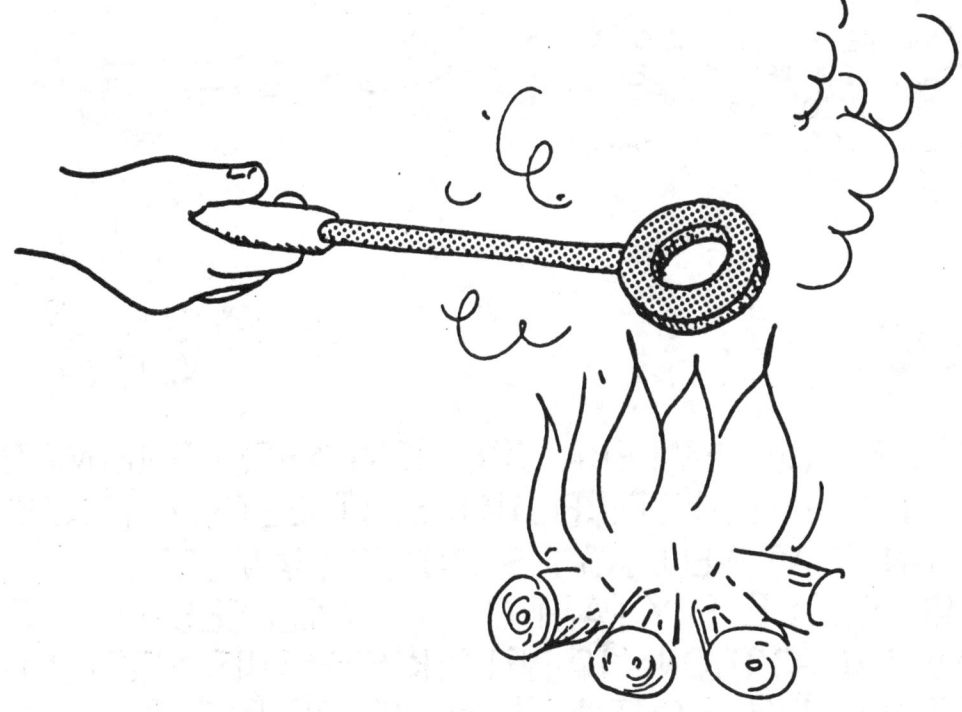

WHEN THE TEMPERATURE OF A METAL RING INCREASES, DOES THE HOLE BECOME LARGER? SMALLER? OR STAY THE SAME SIZE?

CHAP 14

CONCEPTUAL Physics

WHEN THE TEMPERATURE OF A METAL RING INCREASES, DOES THE HOLE BECOME LARGER? SMALLER? OR STAY THE SAME SIZE?

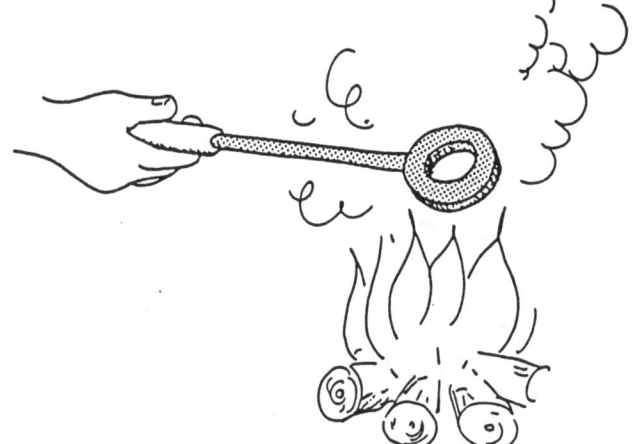

ANSWER:

WHEN THE TEMPERATURE INCREASES, THE METAL EXPANDS---IN ALL DIRECTIONS. IT GETS THICKER; ITS INNER AS WELL AS ITS OUTER DIAMETER INCREASES; EVERY PART OF IT INCREASES BY THE SAME PROPORTION. TO BETTER SEE THIS, PRETEND THAT THE RING IS CUT IN FOUR PIECES BEFORE BEING HEATED. WHEN HEATED THEY ALL EXPAND. CAN YOU SEE WHEN THEY ARE REASSEMBLED THAT THE HOLE IS LARGER?

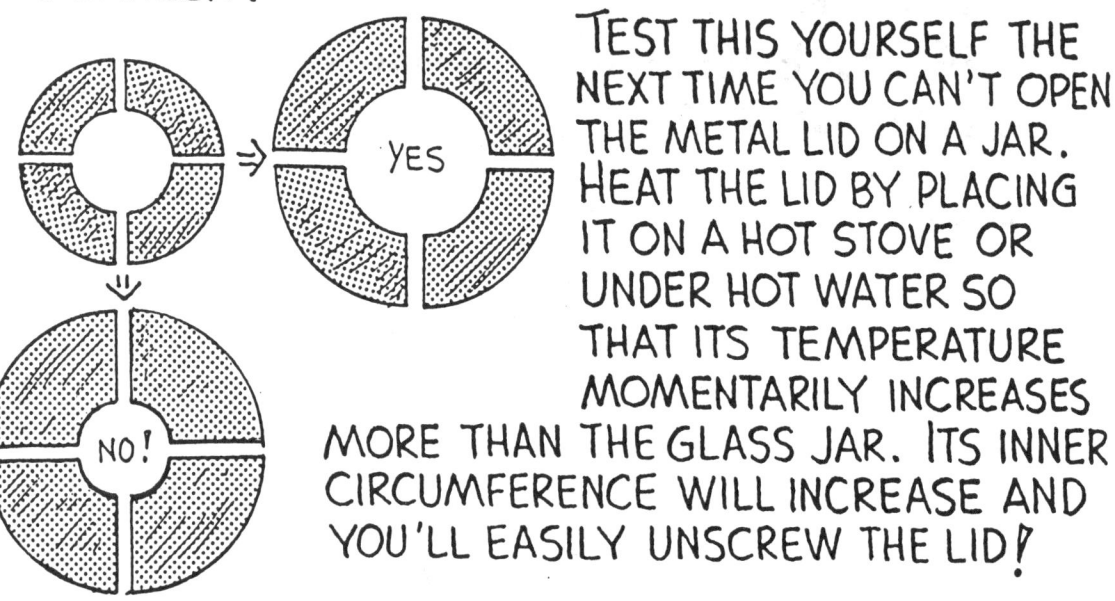

TEST THIS YOURSELF THE NEXT TIME YOU CAN'T OPEN THE METAL LID ON A JAR. HEAT THE LID BY PLACING IT ON A HOT STOVE OR UNDER HOT WATER SO THAT ITS TEMPERATURE MOMENTARILY INCREASES MORE THAN THE GLASS JAR. ITS INNER CIRCUMFERENCE WILL INCREASE AND YOU'LL EASILY UNSCREW THE LID!

CONCEPTUAL Physics

When the temperature of the piece of metal is increased and the metal expands, will the gap between the ends become narrower, or wider, or remain unchanged?

CONCEPTUAL Physics

When the temperature of the piece of metal is increased and the metal expands, will the gap between the ends become narrower, or wider, or remain unchanged?

ANSWER:

The gap will become wider when the metal expands. To see this, pretend the shape is composed of little blocks, each the size of the gap. When heated, each block expands the same. So if the metal is heated uniformly, every part expands at the same rate ⋯ even the gap.

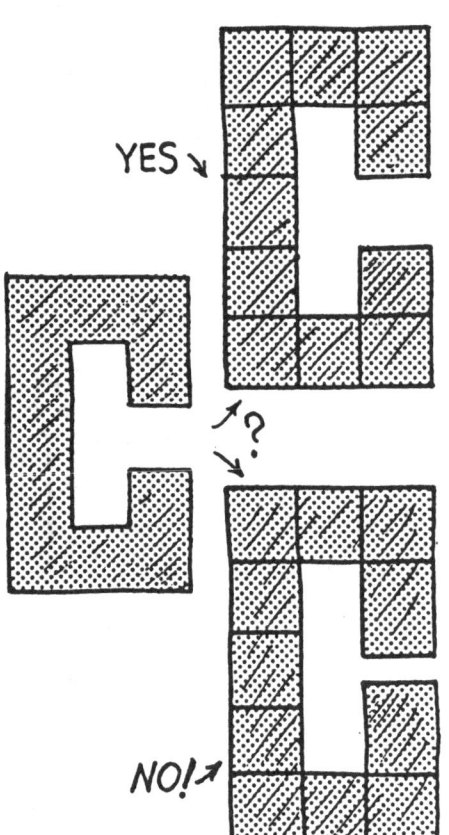

CHAP. 14

CONCEPTUAL **Physics**

Touch the inside of a 200°C hot oven and you burn yourself. But when the 1200°C white hot sparks from a 4th-of-July-type sparkler hit your skin, you're okay. Why?

CONCEPTUAL Physics

Touch the inside of a 200°C hot oven and you burn yourself. But when the 1200°C white hot sparks from a 4th-of-July-type sparkler hit your skin, you're okay. Why?

Answer:

Temperature is proportional to energy per molecule. How much energy depends on how many molecules. When you touch the inside surface of the oven, you're making contact with many, many molecules, and the flow of energy is a painful experience. Although the energy per molecule is much greater in the sparks of the firework, you make contact with only a relatively few molecules when a spark lands on you. The corresponding low energy transfer borders on your threshold of feeling.

> High temperature at low energy is like high voltage at low energy. Both a high-temperature spark and the high voltage of a charged balloon are harmless because their energies are very small.

Hewitt Drewit!

CONCEPTUAL Physics

Suppose in a restaurant your coffee is served about 5 or 10 minutes before you are ready for it. In order that it be as hot as possible when you drink it, should you pour in the room-temperature cream right away or when you are ready to drink the coffee?

CONCEPTUAL Physics

Suppose in a restaurant your coffee is served about 5 or 10 minutes before you are ready for it. In order that it be as hot as possible when you drink it, should you pour in the room-temperature cream right away or when you are ready to drink the coffee?

ANSWER:

Pour the cream in right away. In so doing, you lighten the color of the coffee. When the coffee is black, it is a better radiator and will cool faster than when it is lighter in color. Perhaps you can think of some other reasons for pouring the cream right away.

CHAP. 15

CONCEPTUAL Physics

WHAT IS THE MINIMUM AMOUNT OF 100°C STEAM REQUIRED TO MELT 1 GRAM OF 0°C ICE?

ANSWER:

0.125 GRAM OF 100°C STEAM WILL PROVIDE THE 80 CALORIES REQUIRED TO MELT 1 GRAM OF ICE. THE H_2O IN THE FORM OF STEAM WILL GIVE UP 540 CALORIES PER GRAM WHEN IT CONDENSES TO BOILING WATER, AND ANOTHER 100 CALORIES PER GRAM WHEN THE WATER IS COOLED FROM 100°C TO 0°C. SO THE STEAM WILL GIVE UP A TOTAL OF 640 CALORIES PER GRAM TO THE ICE. BUT THE ICE NEEDS ONLY 80 CALORIES TO MELT. SO ONLY 80/640 GRAM (0.125 GRAM) OF STEAM WILL DO THE JOB.

CONCEPTUAL Physics

Suppose 4 grams of boiling water are spread over a large surface so 1 gram rapidly evaporates. If evaporation takes 540 calories from the remaining 3 grams of water, and no other heat transfer takes place, what will be the temperature of the remaining 3 grams?

CHAP. 16

CONCEPTUAL Physics

Suppose 4 grams of boiling water are spread over a large surface so 1 gram rapidly evaporates. If evaporation takes 540 calories from the remaining 3 grams of water, and no other heat transfer takes place, what will be the temperature of the remaining 3 grams?

ANSWER:

The remaining 3 grams will turn to 0°C ice under conditions where all 540 calories are taken from the remaining water (like when the surroundings are below freezing and don't contribute energy). 540 calories from 3 grams means each gram gives up 180 calories. 100 calories from a gram of boiling water reduces its temperature to 0°C, and 80 more calories taken away turns it to ice. This is why hot water so quickly turns to ice in a freezing-cold environment.

CHAP. 16

CONCEPTUAL Physics

He can quickly walk barefoot across red hot coals of wood without harm because of

a) mind of matter

b) reasons that are outside mainstream physics

c) basic physics concepts

CONCEPTUAL Physics

He can quickly walk barefoot across red hot coals of wood without harm because of

a) mind of matter
b) reasons that are outside mainstream physics
c) basic physics concepts

The answer is c:

First of all, the coals are wood, a very poor conductor of heat. Wood is a poor conductor even when it's hot, which is why wooden handles are used on cookware. Even when the wood is red hot, its poor conductivity allows quick steps without the transfer of very much heat. High temperature and how much heat transfers are entirely different physics concepts. Secondly, if your feet are damp because of perspiration or wet surrounding grass, even less heat is transferred to your feet. Why? Two reasons: Some of the heat energy goes into evaporating the moisture that would otherwise burn you — and when the moisture turns to vapor it provides an insulating blanket. This is why you wet your finger before touching a hot clothes iron.

For mind-over-matter advocates, try walking on red hot coals of iron — ouch!

Caution: Walking on red hot coals is very dangerous and many people have accidentally burned themselves.

CHAP. 16

CONCEPTUAL **Physics**

The efficiency of a common incandescent lamp for converting electrical energy into heat is about

- a)   5 %
- b) 20 %
- c) 100 %

CHAP. 17

CONCEPTUAL Physics

The efficiency of a common incandescent lamp for converting electrical energy into heat is about

a)   5 %
b) 20 %
c) 100 %

The answer is C, 100% :

Although its efficiency for converting electrical energy into light is about 5%, all the energy dissipated by the lamp, even that momentarily converted to light, becomes heat.

That's why it isn't wasteful to keep the lights on in a building that is being electrically heated!

CONCEPTUAL Physics

The air temperature at an altitude of 10 kilometers is a chilling -35°C. Cabin temperatures in airplanes flying at this altitude are comfortable because of air conditioners rather than heaters. Why?

CHAP. 17

CONCEPTUAL Physics

The air temperature at an altitude of 10 kilometers is a chilling −35°C. Cabin temperatures in airplanes flying at this altitude are comfortable because of air conditioners rather than heaters. Why?

ANSWER:

Airliners have pressurized cabins. The process of stopping and compressing outside air to near-sea-level pressures would normally heat the air to a roasting 55°C (130°F).

So air conditioners must be used to extract heat from the pressurized air.

CONCEPTUAL Physics

A PIECE OF IRON HAS A TEMPERATURE OF 0°C. A SECOND IDENTICAL PIECE OF IRON IS TWICE AS HOT. WHAT IS THE TEMPERATURE OF THE SECOND PIECE OF IRON?

CHAP. 17

CONCEPTUAL Physics

A PIECE OF IRON HAS A TEMPERATURE OF 0°C. A SECOND IDENTICAL PIECE OF IRON IS TWICE AS HOT. WHAT IS THE TEMPERATURE OF THE SECOND PIECE OF IRON?

ANSWER:

THE TWICE-AS-HOT IRON HAS A TEMPERATURE OF 273°C. WHY? BECAUSE IT HAS TWICE THE THERMAL ENERGY AND WILL HAVE TWICE THE ABSOLUTE TEMPERATURE. IF THE IRON HAD NO THERMAL ENERGY ITS TEMPERATURE WOULD BE ABSOLUTE ZERO -- THAT'S -273°C. SO TWICE AS HOT MUST BE +273°C.

CONCEPTUAL Physics

A PIECE OF IRON HAS A TEMPERATURE OF 10°C. A SECOND IDENTICAL PIECE OF IRON IS TWICE AS HOT. WHAT IS THE TEMPERATURE OF THE SECOND PIECE OF IRON?

ANSWER:

THE TWICE-AS-HOT IRON IS 293°C:

CONSIDER A STICK THAT IS 273 + 10 UNITS LONG. THIS IS LIKE A THERMOMETER THAT EXTENDS FROM ABSOLUTE ZERO (-273°C) TO 10°C. CAN YOU SEE THAT A STICK TWICE AS LONG IS 2 × 283 = 566 UNITS LONG? (OR TEMPERATURE-WISE, 566 K?)

SUBTRACT THE 273 PART AND YOU HAVE 566 - 273 = 293 UNITS --- LIKEWISE FOR THE TWICE-AS-HOT 10°C IRON.

CHAP. 17

CONCEPTUAL Physics

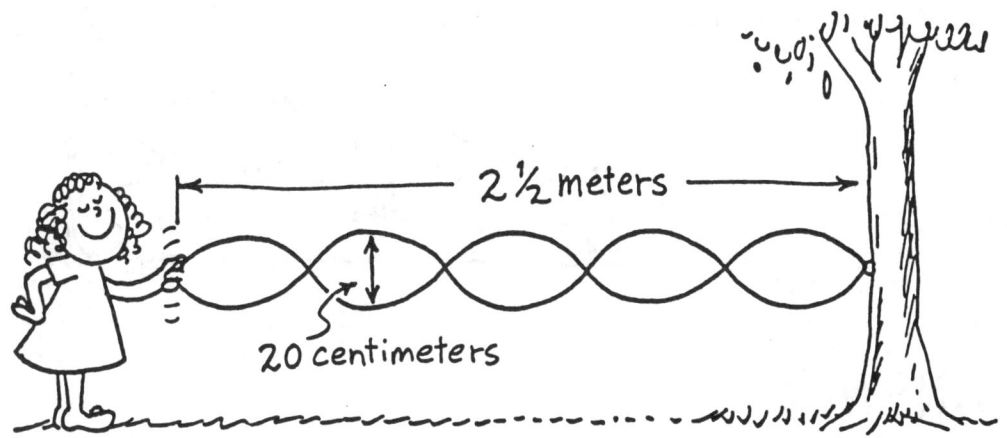

IN THE STANDING WAVE SHOWN,
WHAT IS ITS AMPLITUDE?
WHAT IS ITS WAVELENGTH?
HOW MANY NODES ARE THERE?

CONCEPTUAL Physics

IN THE STANDING WAVE SHOWN,
WHAT IS ITS AMPLITUDE?
WHAT IS ITS WAVELENGTH?
HOW MANY NODES ARE THERE?

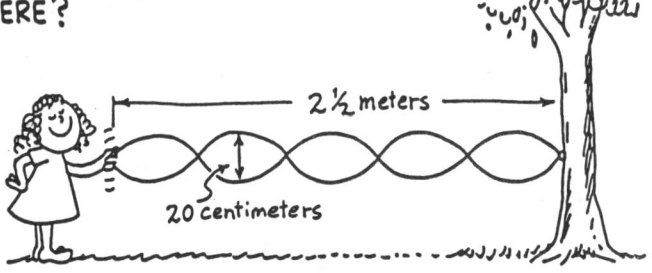

ANSWER:

THE AMPLITUDE OF THE WAVE IS 10 CENTIMETERS; THE WAVELENGTH IS 1 METER; AND THERE ARE 6 NODES.

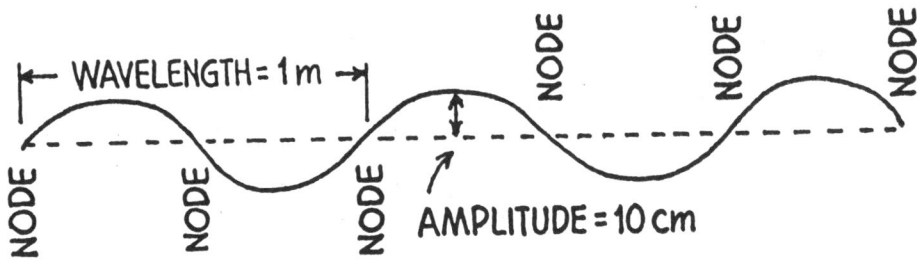

CHAP. 18

CONCEPTUAL Physics

A CONICAL SHOCK WAVE IS GENERATED BY A SUPERSONIC AIRCRAFT AS SHOWN.

ESTIMATE THE SPEED OF THE CRAFT.

CHAP. 18

CONCEPTUAL Physics

A CONICAL SHOCK WAVE IS GENERATED BY A SUPERSONIC AIRCRAFT AS SHOWN.

ESTIMATE THE SPEED OF THE CRAFT.

ANSWER:

TWICE THE SPEED OF SOUND, AS CAN BE SEEN BY THE ANGLE OF THE SHOCK WAVE. EACH SEGMENT OF THE WAVE IS A SUPERPOSITION OF EXPANDING SPHERES THAT WERE GENERATED BY THE CRAFT AS IT TRAVELED ALONG THE DASHED LINE. THE CENTER OF ANY SAMPLE SPHERE (CIRCLE, AS SEEN IN TWO DIMENSIONS ON THE PAGE) SHOWS WHERE THE CRAFT WAS WHEN THE SPHERE WAS FIRST PRODUCED. BY COMPARING THE DISTANCE THE SOUND HAS TRAVELED TO THE DISTANCE THE CRAFT HAS TRAVELED IN THE SAME TIME, WE HAVE THE SPEED OF THE CRAFT COMPARED TO THE SPEED OF SOUND. IN THIS CASE, THE CRAFT HAS TRAVELED TWICE AS FAR AS SOUND HAS TRAVELED IN THE SAME TIME, SO IT MOVES AT TWICE THE SPEED OF SOUND.

CONCEPTUAL Physics

Suppose at a concert a singer's voice is radio broadcast all the way around the world before reaching the radio you hold to your ear. This takes 1/8 second. If you're close, you hear her voice in air before you hear it from the radio. But if you are far enough away, both signals will reach you at the same time. How many meters distant must you be for this to occur?

CONCEPTUAL Physics

SUPPOSE AT A CONCERT A SINGER'S VOICE IS RADIO BROADCAST ALL THE WAY AROUND THE WORLD BEFORE REACHING THE RADIO YOU HOLD TO YOUR EAR. THIS TAKES 1/8 SECOND. IF YOU'RE CLOSE, YOU HEAR HER VOICE IN AIR BEFORE YOU HEAR IT FROM THE RADIO. BUT IF YOU ARE FAR ENOUGH AWAY, BOTH SIGNALS WILL REACH YOU AT THE SAME TIME. HOW MANY METERS DISTANT MUST YOU BE FOR THIS TO OCCUR?

ANSWER:

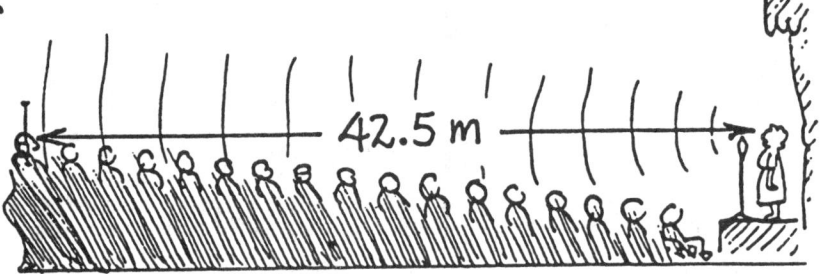

IF YOU SIT 42.5 METERS AWAY FROM THE SINGER, BOTH THE SOUND FROM THE RADIO THAT IS BROADCAST ALL THE WAY AROUND THE WORLD AND THAT THROUGH THE AIR WILL REACH YOU IN THE SAME 1/8 SECOND.

DISTANCE IN AIR = SPEED OF SOUND × TIME IN AIR
= 340 m/s × 1/8 s
= 42.5 m

IF YOU SIT FARTHER BACK, YOU'LL HEAR THE RADIO SIGNAL BEFORE YOU HEAR THE SOUND SIGNAL!

CHAP. 19

CONCEPTUAL Physics

Does the wind affect the pitch of the factory whistle you hear on a windy day?

If so, why?

If not, why not?

CHAP. 20

CONCEPTUAL Physics

Does the wind affect the pitch of the factory whistle you hear on a windy day?

If so, why?

If not, why not?

ANSWER:

No, the wind does not affect the pitch. The wind does affect the speed of sound because the medium that carries the sound moves. But the wavelength of the sound changes accordingly, which results in no change in frequency or pitch. This can be seen by analogy:

Suppose a friend is placing packages on a conveyor belt, say at a "frequency" of one each second. Then you, at the other end of the belt, take off one package each second. Suppose the speed of the belt increases while your friend still places one package per second on the belt. Can you see that the packages (farther apart now) will still arrive to you at the rate of one per second?

Which is more dangerous, touching a faulty 110-volt light bulb or a Van de Graaff generator charged to 100,000 volts? Why?

CONCEPTUAL Physics

Which is more dangerous, touching a faulty 110-volt light bulb or a Van de Graaff generator charged to 100,000 volts? Why?

Answer:

Touching the Van de Graaff generator may be a hair-raising experience, but touching the 110-volt faulty fixture could be the last thing you do. The charged generator nicely illustrates the difference between electric potential energy and electric potential. Electric potential is electric potential energy *per charge*. Although the generator may be charged to an electric potential of 100,000 volts, the amount of charge is relatively small. That and the short time of charge transfer is why you're normally not harmed when it discharges through you. In contrast, if you become the shortcircuit for household 110 volts, the sustained transfer of charge is appreciable. Less energy per charge, but many, many more charges!

$\frac{E}{q} = 110$ V

$\frac{E}{q} = 100,000$ V

CHAP. 21

CONCEPTUAL Physics

ARE OCCUPANTS OF AN AIRPLANE FLYING IN THE MIDST OF A THUNDERSTORM IN DANGER OF BEING STRUCK BY LIGHTNING?

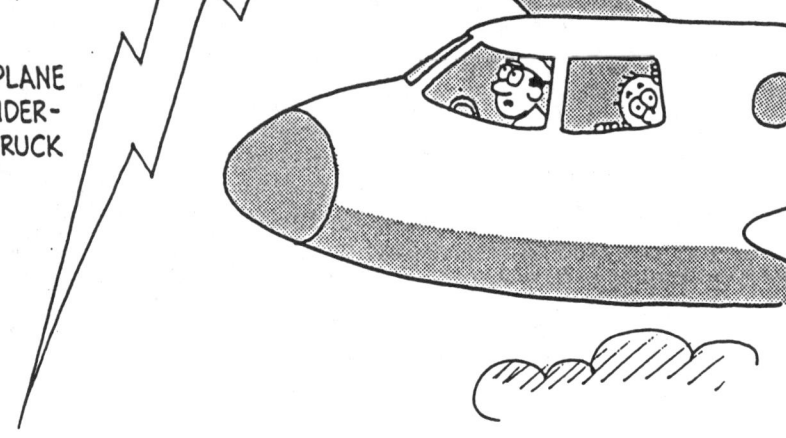

ANSWER:

NO. SINCE THE PLANE DOES NOT OFFER A CONDUCTING PATH TO GROUND, IT IS UNLIKELY THAT IT WOULD BE STRUCK BY LIGHTNING. IF IT IS STRUCK, ELECTRIC CHARGES WILL MUTUALLY REPEL ONE ANOTHER AND NOT PENETRATE THE METAL SURFACE, WHICH ACTS AS AN ELECTRICAL SHIELD.

CONCEPTUAL Physics

Two oppositely charged particles, an alpha particle with two positive charges and a less-massive electron with a single negative charge are attracted to each other. Compared to the force that the alpha particle exerts on the electron, the electron exerts a force on the alpha particle that is

    a) GREATER
    b) THE SAME
    c) LESS

The particle with the most acceleration is the

    d) ALPHA PARTICLE
    e) ELECTRON
    f) SAME FOR EACH

As the particles get closer to each other, each experiences an <u>INCREASE</u> in

    g) FORCE
    h) SPEED
    i) ACCELERATION
    j) ALL OF THESE
    k) NONE OF THESE

CHAP. 21

CONCEPTUAL Physics

Two oppositely charged particles, an alpha particle with two positive charges and a less-massive electron with a single negative charge are attracted to each other. Compared to the force that the alpha particle exerts on the electron, the electron exerts a force on the alpha particle that is

    a) greater
    b) the same
    c) less

The particle with the most acceleration is the

    d) alpha particle
    e) electron
    f) same for each

As the particles get closer to each other, each experiences an <u>increase</u> in

    g) force
    h) speed
    i) acceleration
    j) all of these
    k) none of these

The answers are b, e, and j:

By Newton's 3rd law, the particles pull on each other with equal and opposite forces. By Newton's 2nd law, for the same force the particle with less mass undergoes more acceleration. By Coulomb's law, as the separation distance is decreased, the force increases. By Newton's 2nd law, as the force increases the acceleration increases. Since the particles accelerate toward each other, their speeds increase also.

CONCEPTUAL Physics

Touch the terminals of a 100-volt battery and you're jolted. Touch a 10,000-volt rubber balloon and you feel nothing. Why?

CAREFUL: TOUCHING HIGH-VOLTAGE TERMINALS IS A SAFETY NO NO!

CONCEPTUAL Physics

Touch the terminals of a 100-volt battery and you're jolted. Touch a 10,000-volt rubber balloon and you feel nothing. Why?

CAREFUL: TOUCHING HIGH-VOLTAGE TERMINALS IS A SAFETY NO NO!

Answer:

Voltage is energy per charge. How much energy depends on how much charge. A lot more charge flows through you when you become the circuit for the battery, and the corresponding energy flow can be a shocking experience. Although the energy per charge is 100 times greater on the balloon, only about a millionth as much charge flows through you when you discharge it. The corresponding low energy flow is below your threshold of feeling.

If as much charge flowed through you in touching the high-voltage balloon as the low-voltage battery, you'd be in trouble!

High voltage at low energy is like high temperature at low energy. Both a high-voltage balloon and the high-temperature white-hot sparks of a 4th-of-July sparkler are harmless because their energies are very small.

CONCEPTUAL Physics

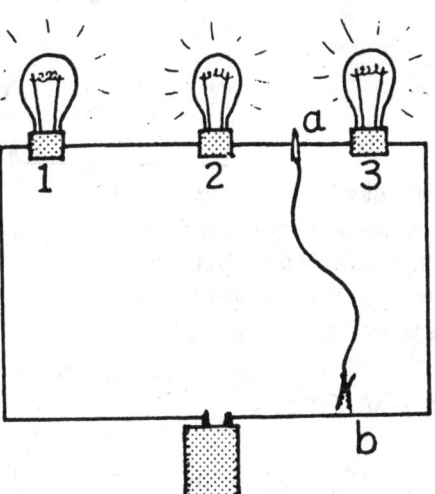

The simple series circuit consists of three identical lamps powered by battery. When a wire is connected between points a and b,

a) what happens to the brightness of lamp 3?

b) does current in the circuit increase, decrease, or remain the same?

c) what happens to the brightness of lamps 1 and 2?

d) does the voltage drop across lamps 1 and 2 increase, decrease, or remain the same?

e) is the power dissipated by the circuit increased, decreased, or does it remain the same?

CONCEPTUAL Physics

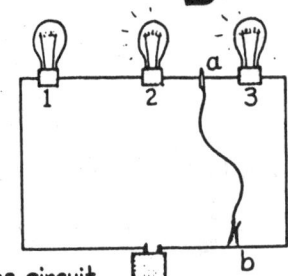

The simple series circuit consists of three identical lamps powered by battery. When a wire is connected between points *a* and *b*,

a) what happens to the brightness of lamp 3?

b) does current in the circuit increase, decrease, or remain the same?

c) what happens to the brightness of lamps 1 and 2?

d) does the voltage drop across lamps 1 and 2 increase, decrease, or remain the same?

e) is the power dissipated by the circuit increased, decreased, or does it remain the same?

Answers:

a) Lamp 3 is short-circuited. It no longer glows because no current passes through it.

b) The current in the circuit increases. Why? Because the circuit resistance is reduced. Whereas charge was made to flow through three lamps before, now it flows through only two lamps -- 2/3 the resistance results in 3/2 the current (neglecting temperature effects).

c) Lamps 1 and 2 glow brighter because of the increased current through them.

d) The voltage drop across lamps 1 and 2 is greater. Whereas voltage supplied by the battery was previously divided between three lamps, it is now divided between only two lamps. So more energy is now given to each lamp.

e) The power output of the two-lamp circuit is greater because of the greater current. This means more light will be emitted by the two lamps in series than from the three lamps in series. Three lamps connected in parallel, however, put out more light. Lamps are most often connected in parallel.

Hewitt Drewit!

CONCEPTUAL Physics

Which circuit draws the most current?

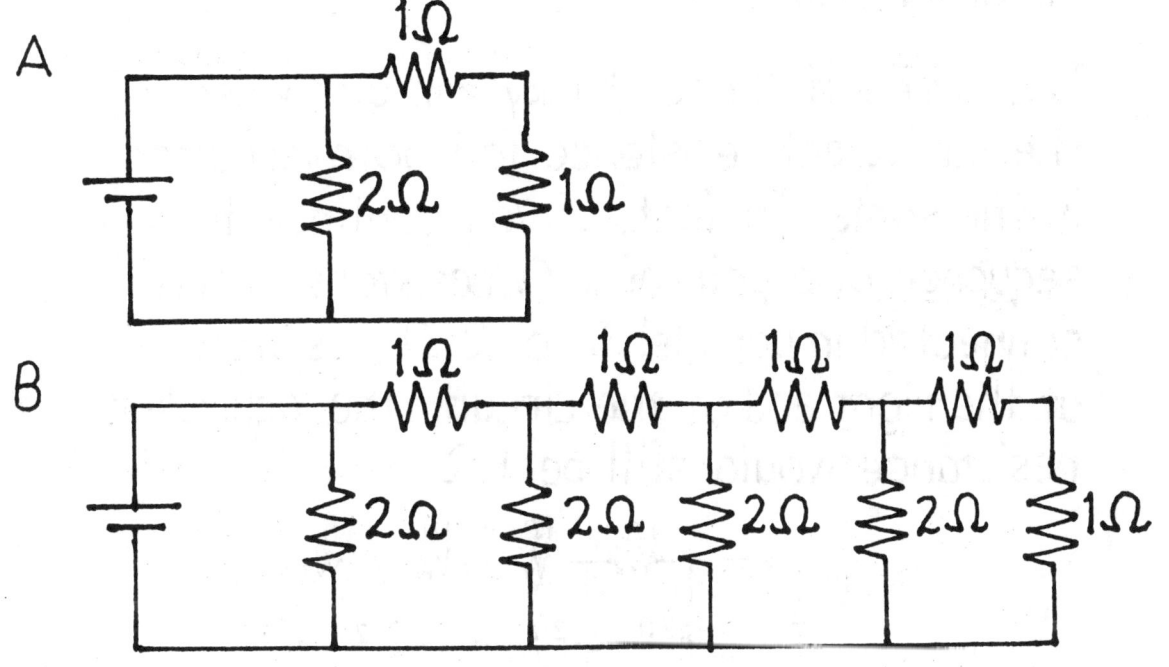

a) Circuit A

b) Circuit B

c) Both the same

CONCEPTUAL Physics

Which circuit draws the most current?

a) Circuit A
b) Circuit B
c) Both the same

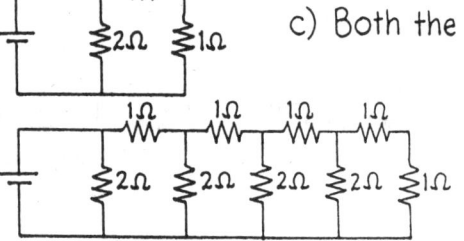

The answer is c:

This is one of those "tricky" circuits wherein the equivalent resistance for both circuits is the same. In fact, if you continue the sequence of a pair of 1-Ω resistors in series connected in parallel to a 2-Ω resistor at the right end of the circuit, the equivalent resistance would still be 1 Ω.

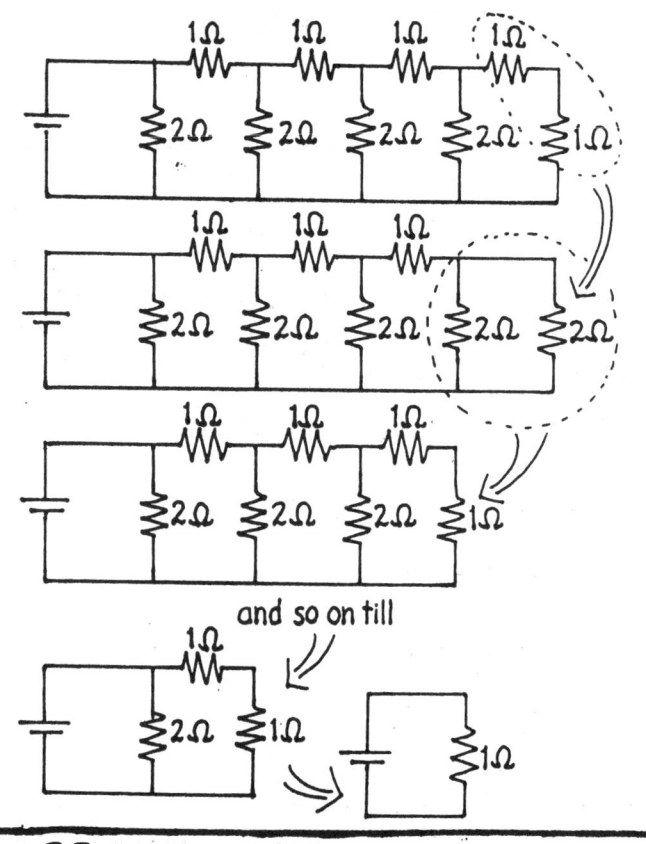

and so on till

CHAP. 22

THE 40-WATT BULB AND THE 100-WATT BULB ARE CONNECTED IN SERIES TO THE BATTERY. WHICH BULB WILL GLOW BRIGHTER?

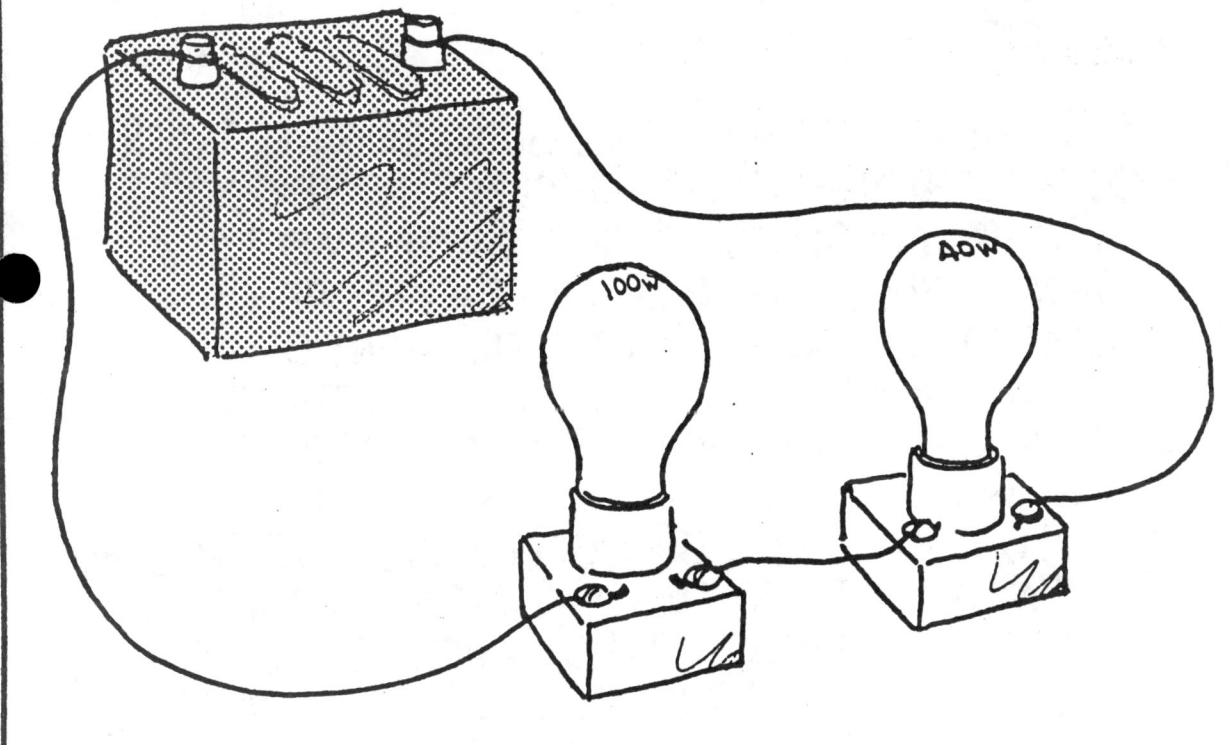

CONCEPTUAL Physics

THE 40-WATT BULB AND THE 100-WATT BULB ARE CONNECTED IN SERIES TO THE BATTERY. WHICH BULB WILL GLOW BRIGHTER?

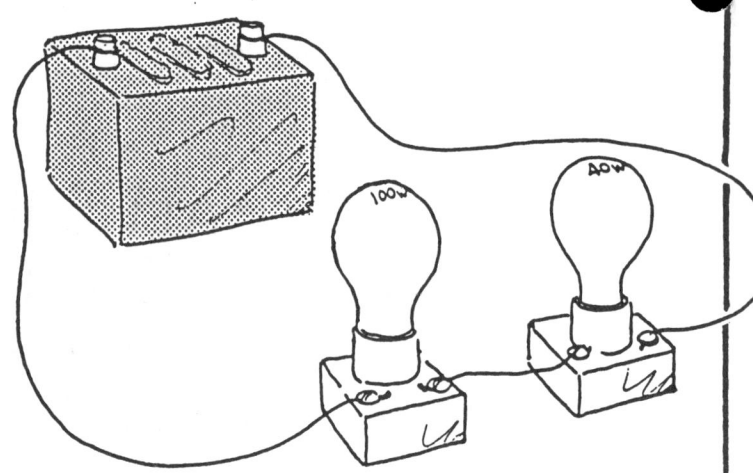

ANSWER:

THE 40-WATT BULB WILL GLOW BRIGHTER WHEN CONNECTED IN SERIES. TO UNDERSTAND THIS YOU MUST FIRST UNDERSTAND THAT THE FILAMENT IN A 40-WATT BULB IS THINNER AND THEREFORE OF HIGHER RESISTANCE THAN THE FILAMENT OF A 100-WATT BULB. IT IS THE HIGHER RESISTANCE OF THE 40-WATT BULB THAT KEEPS THE CURRENT TO ONLY 40/100 THE CURRENT IN A 100-WATT BULB WHEN BOTH ARE PROPERLY CONNECTED IN *PARALLEL*. THEN MORE CURRENT FLOWS IN THE 100-WATT BULB AND IT GLOWS BRIGHTER. BUT WHEN CONNECTED IN SERIES, THE CURRENT IS LESS BUT IS THE SAME IN EACH. THE SAME AMOUNT OF CURRENT "SQUEEZING" THROUGH THE FINER FILAMENT OF THE 40-WATT BULB HEATS IT MORE AND MAKES IT GLOW BRIGHTER THAN THE LOWER-RESISTANCE 100-WATT BULB.

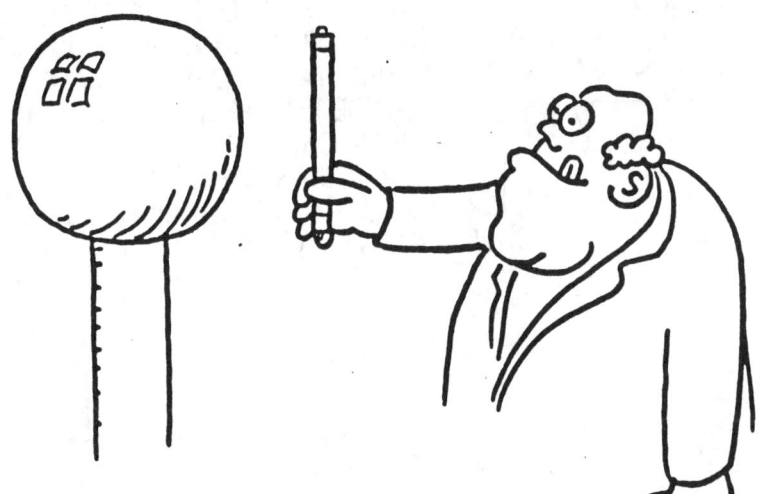

THE LAMP WILL NOT GLOW WHEN IT IS HELD WITH BOTH ENDS EQUIDISTANT FROM THE CHARGED VAN DE GRAAFF GENERATOR. BUT WHEN ONE END IS CLOSER TO THE DOME THAN THE OTHER END, A CURRENT IS ESTABLISHED AND IT GLOWS. WHY?

CONCEPTUAL Physics

THE LAMP WILL NOT GLOW WHEN IT IS HELD WITH BOTH ENDS EQUIDISTANT FROM THE CHARGED VAN DE GRAAFF GENERATOR. BUT WHEN ONE END IS CLOSER TO THE DOME THAN THE OTHER END, A CURRENT IS ESTABLISHED AND IT GLOWS. WHY?

ANSWER:

SIMPLY PUT, THE END OF THE LAMP THAT IS HELD CLOSER TO THE DOME IS AT A HIGHER ELECTRIC POTENTIAL THAN THE FARTHER END. THE ELECTRIC POTENTIAL DIFFERENCE ACROSS THE ENDS OF THE LAMP PRODUCES A CURRENT IN THE LAMP.

MORE ACCURATELY, CHARGE ARCS FROM THE DOME THROUGH THE AIR TO THE CLOSEST PART OF THE LAMP, THEN THROUGH THE LAMP TO YOUR HAND, AND THEN THROUGH YOUR BODY TO THE FLOOR AND BACK TO THE GENERATOR TO FORM A CONTINUOUS LOOP. WHEN THE LAMP IS HELD IN THE FIRST POSITION, BOTH ENDS ARE EQUIDISTANT AND CHARGE DOES NOT FLOW THROUGH THE LENGTH OF THE LAMP TO COMPLETE THE CIRCUIT.

CONCEPTUAL Physics

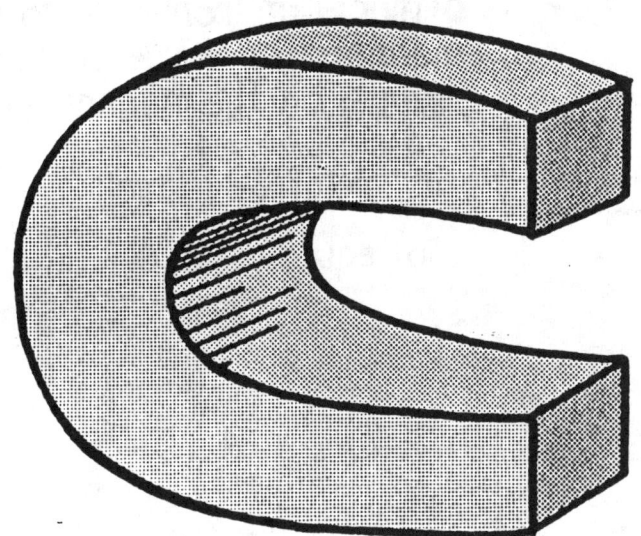

Compared to the huge force that attracts an iron tack to a strong magnet, the force that the tack exerts on the magnet is

a) relatively small

b) equally huge

CONCEPTUAL Physics

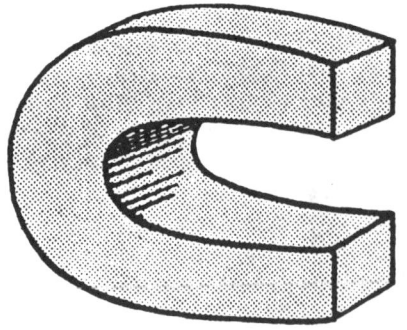

Compared to the huge force that attracts an iron tack to a strong magnet, the force that the tack exerts on the magnet is

a) relatively small
b) equally huge

Answer:

The pair of forces between the tack and magnet comprises a single interaction and both are equal in magnitude and opposite in direction—Newton's third law.

Like which pulls harder on the stretched rubber band — my thumb or my finger?

CHAP. 23

CONCEPTUAL Physics

When current flows in the wire that is placed in the magnetic field shown, the wire is forced upward. If the wire is made to form a loop as shown below, the loop will tend to

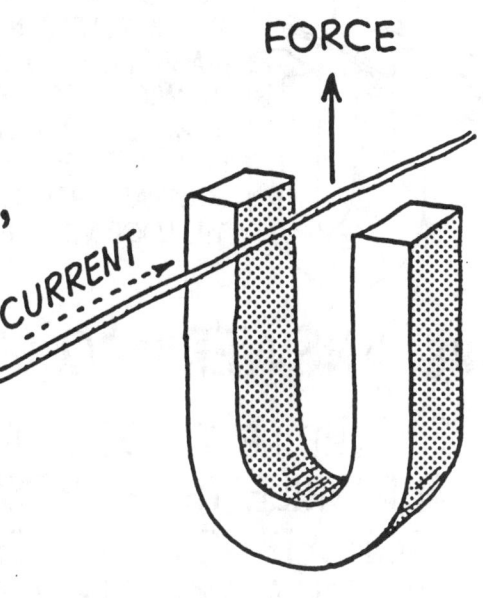

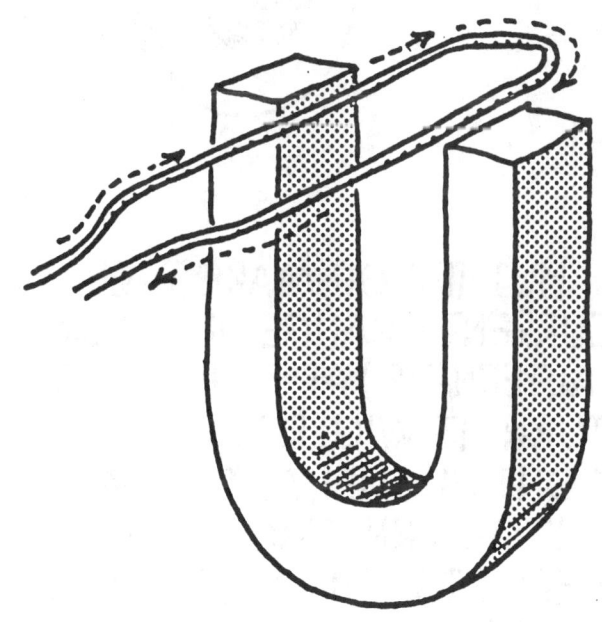

a) ROTATE CLOCKWISE

b) ROTATE COUNTER-CLOCKWISE

c) REMAIN AT REST

CONCEPTUAL Physics

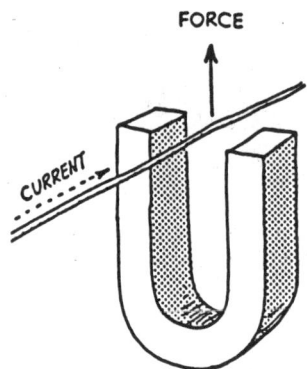

WHEN CURRENT FLOWS IN THE WIRE THAT IS PLACED IN THE MAGNETIC FIELD SHOWN THE WIRE IS FORCED UPWARD. IF THE WIRE IS MADE TO FORM A LOOP AS SHOWN THE LOOP WILL TEND TO

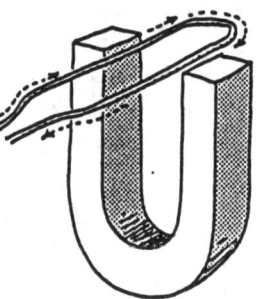

a) ROTATE CLOCKWISE
b) ROTATE COUNTER-CLOCKWISE
c) REMAIN AT REST

THE ANSWER IS a:

THE LEFT SIDE IS FORCED UP WHILE THE RIGHT SIDE IS FORCED DOWN AS SHOWN. IF YOU MAKE THE LOOP ROTATE AGAINST A SPRING AND ATTACH A POINTER TO IT, YOU HAVE A SIMPLE ELECTRIC METER. AT MAXIMUM, IT CAN ONLY MAKE A HALF TURN.

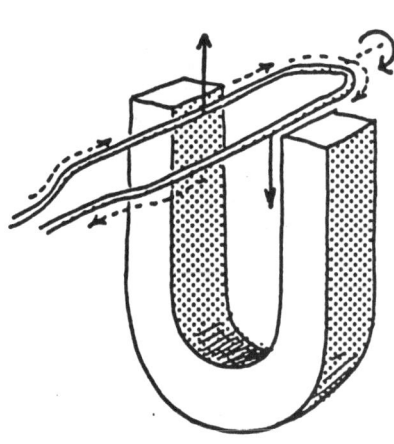

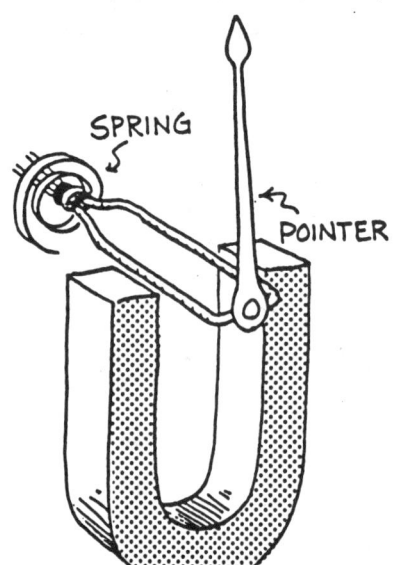

BUT IF YOU MAKE THE CURRENT CHANGE DIRECTION (ALTERNATE) AT EVERY HALF TURN, IT WILL ROTATE CONTINUOUSLY AS LONG AS THE ALTERNATING CURRENT PERSISTS. THEN YOU HAVE A MOTOR.

CHAP. 23

CONCEPTUAL Physics

THE TWO IRON BARS LOOK ALIKE, BUT ONLY ONE IS A MAGNET. HOW CAN YOU DETERMINE WHICH IS THE MAGNET ONLY BY INVESTIGATING THEIR INTERACTION WITH EACH OTHER?

CHAP. 23

CONCEPTUAL Physics

The two iron bars look alike, but only one is a magnet. How can you determine which is the magnet only by investigating their interaction with each other?

SOLUTION:

Make a T shape with the bars. Only when the magnet is placed at the midpoint of the non-magnet will they stick!

CONCEPTUAL Physics

WHAT HAPPENS TO THE READING ON THE GALVANOMETER WHEN THE SWITCH IN CIRCUIT 1 IS a) FIRST CLOSED, b) KEPT CLOSED, AND c) OPENED AGAIN?

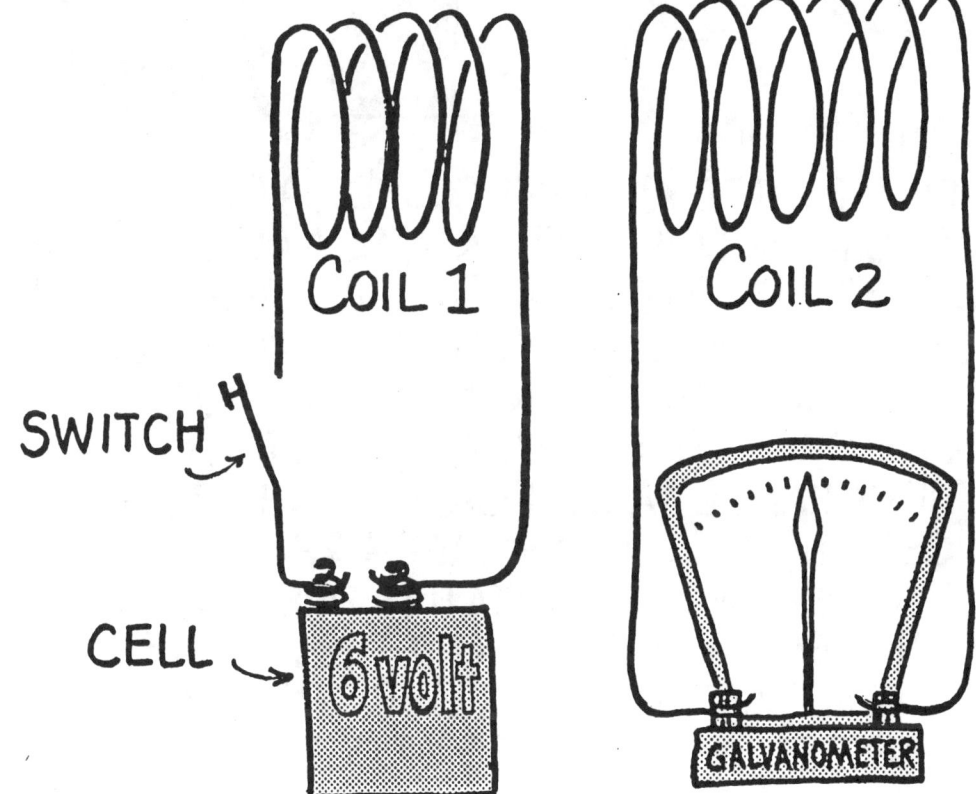

CONCEPTUAL Physics

WHAT HAPPENS TO THE READING ON THE GALVANOMETER WHEN THE SWITCH IN CIRCUIT 1 IS a) FIRST CLOSED, b) KEPT CLOSED, AND c) OPENED AGAIN?

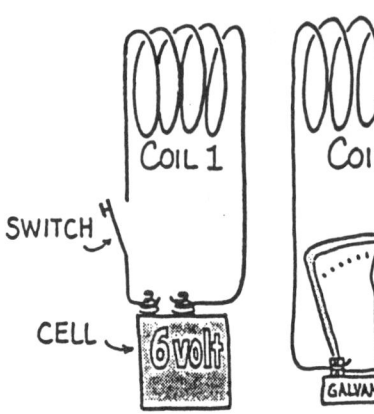
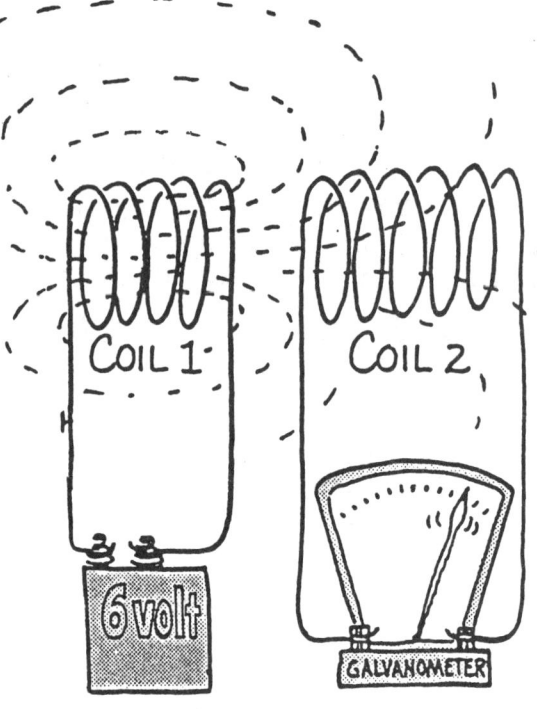

WHEN THE SWITCH IS FIRST CLOSED, A CURRENT IS ESTABLISHED IN COIL 1 AND CREATES A MAGNETIC FIELD WHICH EXTENDS TO COIL 2. THIS BUILD-UP OF FIELD IN COIL 2 INDUCES CURRENT WHICH IS REGISTERED IN THE GALVANOMETER. THE CURRENT IS BRIEF, HOWEVER, BECAUSE ONCE THE FIELD IS STABILIZED AND NO FURTHER CHANGE TAKES PLACE, NO CURRENT IS INDUCED AND THE GALVANOMETER READS ZERO CURRENT. WHEN THE SWITCH IS OPENED, THE CURRENT CEASES IN COIL 1 AND THE MAGNETIC FIELD IN THE COIL AND THE PART THAT EXTENDS TO COIL 2 COLLAPSES. THIS CHANGE INDUCES A PULSE OF CURRENT IN THE OPPOSITE DIRECTION WHICH IS REGISTERED ON THE GALVANOMETER.

Is it correct to say that in every case, without exception, any radio wave travels faster than any sound wave?

CONCEPTUAL Physics

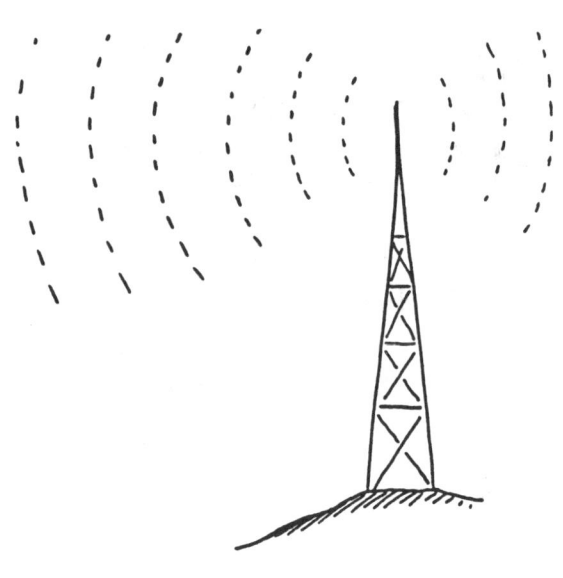

IS IT CORRECT TO SAY THAT IN EVERY CASE, WITHOUT EXCEPTION, ANY RADIO WAVE TRAVELS FASTER THAN ANY SOUND WAVE?

ANSWER:

YES, BECAUSE ANY RADIO WAVE TRAVELS AT THE SPEED OF LIGHT. A RADIO WAVE IS AN ELECTROMAGNETIC WAVE. SO ANY RADIO WAVE, IN A VERY REAL SENSE, IS SIMPLY A LOW-FREQUENCY LIGHT WAVE. A SOUND WAVE, ON THE OTHER HAND, IS FUNDAMENTALLY DIFFERENT. A SOUND WAVE IS A MECHANICAL DISTURBANCE PROPAGATED THROUGH A MATERIAL MEDIUM BY MATERIAL PARTICLES THAT VIBRATE AGAINST ONE ANOTHER. IN AIR, THE SPEED OF SOUND IS ABOUT 340 METERS/SECOND, ABOUT ONE-MILLIONTH THE SPEED OF A RADIO WAVE. SOUND TRAVELS FASTER IN OTHER MEDIA, BUT IN NO CASE AT THE SPEED OF LIGHT. NO SOUND WAVE CAN TRAVEL AS FAST AS LIGHT.

CONCEPTUAL Physics

Openings between leaves in the tree act as "pinholes" and cast images of the sun on the ground below. With only a meter-stick and the knowledge that the sun is 150,000,000 kilometers distant, how can one estimate the sun's diameter?

THANX TO FRANK CRAWFORD

CHAP 25

CONCEPTUAL Physics

Openings between leaves in the tree act as "pinholes" and cast images of the sun on the ground below. With only a meter-stick and the knowledge that the sun is 150,000,000 kilometers distant, how can one estimate the sun's diameter?

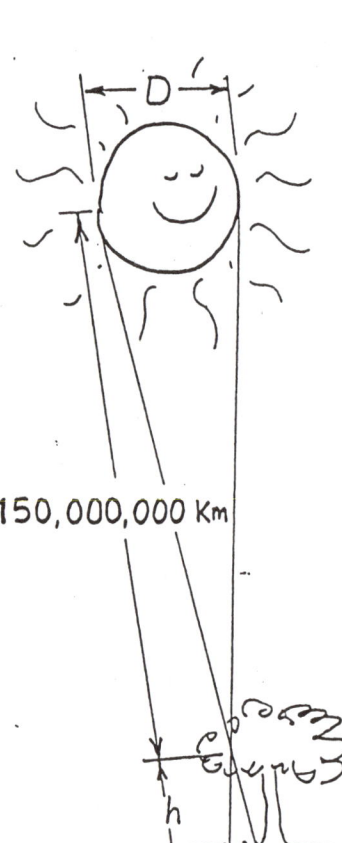

Answer:

Simply measure the diameter d of the sun's image (called a "sunball"), and its distance h below the "pinhole" opening in the leaves. You'll find the ratio $d/h \approx 1/108$, which is also the ratio

$$\frac{\text{Sun's diameter } D}{150,000,000 \text{ km}}$$

So $D = 1/108 \, (150,000,000 \text{ km}) \approx 1,400,000 \text{ km}$.

What shape do sunballs have during a partial eclipse of the sun?

CHAP. 25

CONCEPTUAL Physics

RED   GREEN

IN A DARKENED ROOM IN FRONT OF A WHITE WALL YOU CAST SHADOWS OF YOUR HAND BY TWO COLORED LAMPS, RED AND GREEN, AS SHOWN. THE COLOR OF THE WALL, AND THE LEFT AND RIGHT SHADOWS, WILL RESPECTIVELY APPEAR

a) YELLOW, RED, AND GREEN
b) YELLOW, GREEN, AND RED
c) RED, GREEN, AND YELLOW
d) GREEN, RED, AND YELLOW
e) RED, YELLOW, AND GREEN
f) NONE OF THESE

CHAP. 26

CONCEPTUAL Physics

IN A DARKENED ROOM IN FRONT OF A WHITE WALL YOU CAST SHADOWS OF YOUR HAND BY TWO COLORED LAMPS, RED AND GREEN, AS SHOWN. THE COLOR OF THE WALL, AND THE LEFT AND RIGHT SHADOWS, WILL RESPECTIVELY APPEAR

- a) YELLOW, RED, AND GREEN
- b) YELLOW, GREEN, AND RED
- c) RED, GREEN, AND YELLOW
- d) GREEN, RED, AND YELLOW
- e) RED, YELLOW, AND GREEN
- f) NONE OF THESE

THE ANSWER IS a:

THE BACKGROUND APPEARS YELLOW, THE COMBINATION OF RED AND GREEN. THE LEFT SHADOW IS CAST BY THE GREEN LAMP, AND WOULD APPEAR BLACK IF THE RED LAMP WERE TURNED OFF. BECAUSE LIGHT FROM THE RED LAMP FALLS ON IT, IT IS RED. SIMILARLY, THE RIGHT SHADOW IS CAST BY THE RED LAMP, AND IS ILLUMINATED WITH LIGHT FROM THE GREEN LAMP, SO IT IS GREEN. TRY THIS AND SEE!

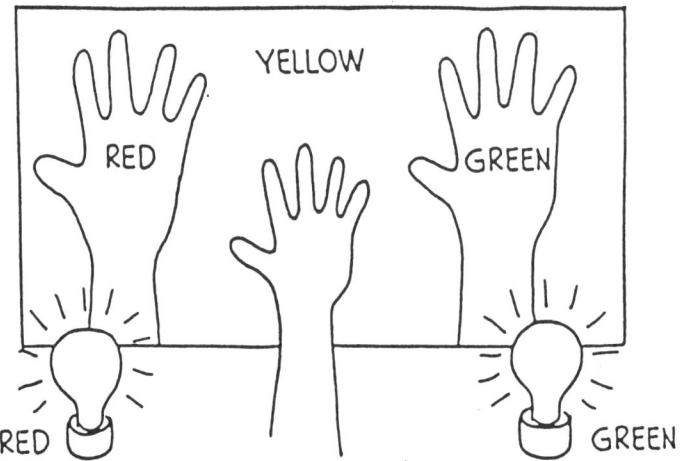

CHAP. 26

CONCEPTUAL Physics

When three colored lamps, red, blue and green, illuminate a physics instructor in front of a white screen in a dark room, three slightly-overlapping shadows appear. Specify the colors in regions 1 through 6.

CONCEPTUAL Physics

When three colored lamps, red, blue and green, illuminate a physics instructor in front of a white screen in a dark room, three slightly-overlapping shadows appear. Specify the colors in regions 1 through 6.

Answers:

Region 1 (shadow of the green lamp) is magenta --- illuminated by red and blue light.

Region 2 (shadow of over-lapped blue and green lamps) is red --- illuminated by only red light.

Region 3 (shadow of the blue lamp) is yellow --- illuminated by red and green light.

Region 4 (shadow of over-lapped red and blue lamps) is green --- illuminated by only green light.

Region 5 (shadow of the red lamp) is cyan --- illuminated by blue and green light.

Region 6 (non-shadowed screen) is white --- the addition of red, green and blue light.

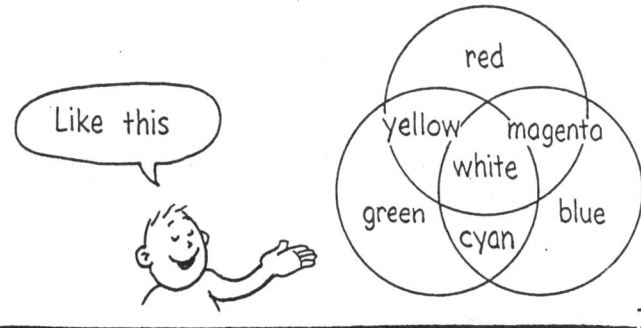

CONCEPTUAL Physics

In order that you are able to see a full-length view of yourself, the minimum size for a plane mirror must be

a) ONE-QUARTER YOUR HEIGHT
b) ONE-HALF YOUR HEIGHT
c) THREE-QUARTERS YOUR HEIGHT
d) YOUR FULL HEIGHT
e) ... DEPENDS ON YOUR DISTANCE

THE ANSWER IS b:

Consistent with the law of reflection, if you look half way down a plane mirror in front of you, you'll see your toes. If you look at parts of the mirror below the half-way mark, you'll see the floor but not yourself. If you look straight ahead, you'll see your eyes. If you look above at a distance half way from your eyes to the top of your head, you'll see the top of your head. You don't see your image in parts of the mirror above. Half way up; half way down --- thats a mirror one-half your height. As the sketch below shows, distance is NOT a factor.

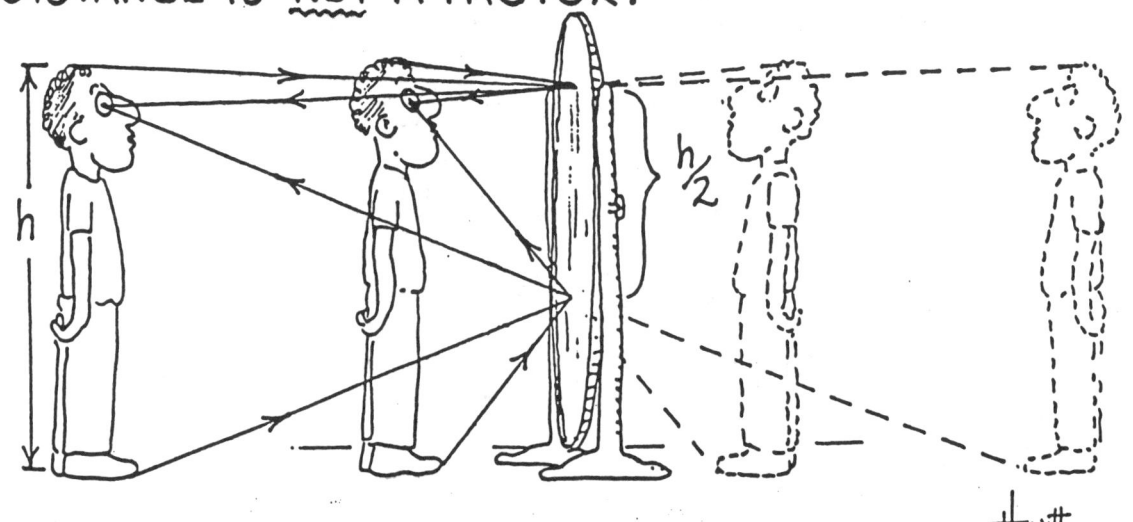

CHAP. 27

CONCEPTUAL Physics

TO SEE MORE OF HER HEAD IN THE MIRROR, SHE

    a) SHOULD HOLD THE MIRROR CLOSER

    b) SHOULD HOLD THE MIRROR FARTHER AWAY

    c) NEEDS A BIGGER MIRROR

CONCEPTUAL Physics

TO SEE MORE OF HER HEAD IN THE MIRROR, SHE

    a) SHOULD HOLD THE MIRROR CLOSER

    b) SHOULD HOLD THE MIRROR FARTHER AWAY

    c) NEEDS A BIGGER MIRROR

THE ANSWER IS C:

IF SHE HOLDS THE MIRROR CLOSER, HER IMAGE APPEARS BIGGER, BUT SO DOES THE MIRROR. IF SHE HOLDS THE MIRROR FARTHER AWAY, BOTH HER IMAGE AND THE MIRROR ARE PROPORTIONALLY REDUCED. AS THE RAY DIAGRAMS SHOW, SHE SEES THE SAME PROPORTION OF HER FACE AT ANY DISTANCE. TRY THIS YOURSELF AND SEE! AND IF YOU CAN NOT SEE YOUR FULL FACE, YOU NEED A BIGGER MIRROR. HOW BIG? AT LEAST HALF THE SIZE OF YOUR FACE.

CHAP. 27

CONCEPTUAL Physics

She stands 1 meter in front of the dresser mirror and looks at the flower on the top of her head in a small mirror held ½ meter behind her head.

How far in back of the dresser mirror does she see the image of the flower?

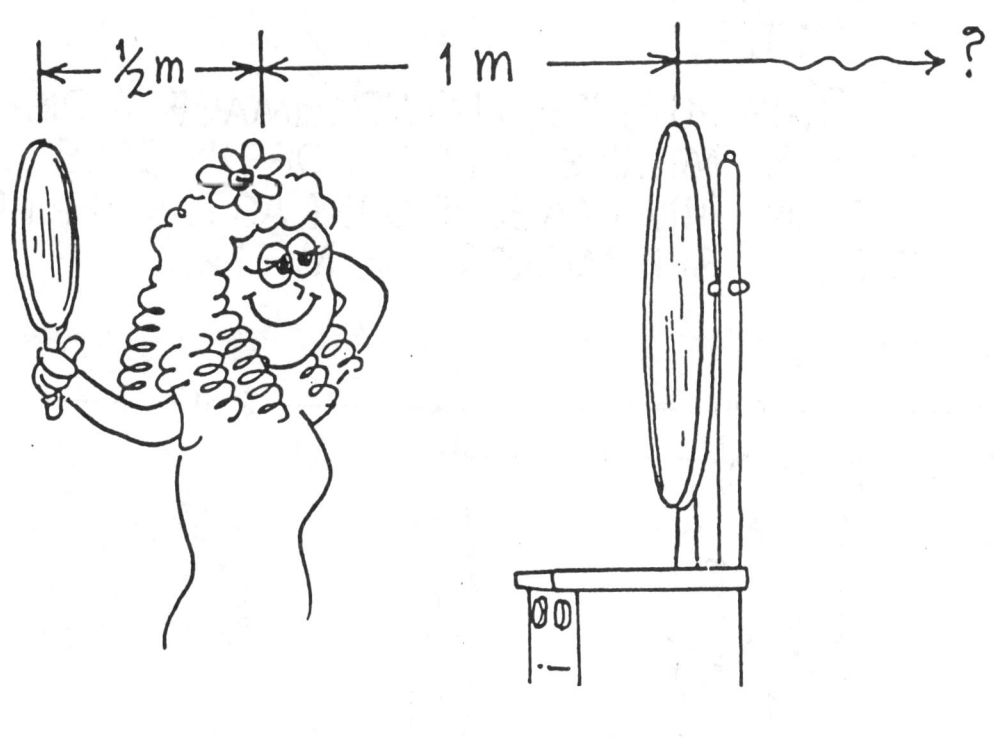

CHAP. 27

CONCEPTUAL Physics

She stands 1 meter in front of the dresser mirror and looks at the flower on the top of her head in a small mirror held ½ meter behind her head.

How far in back of the dresser mirror does she see the image of the flower?

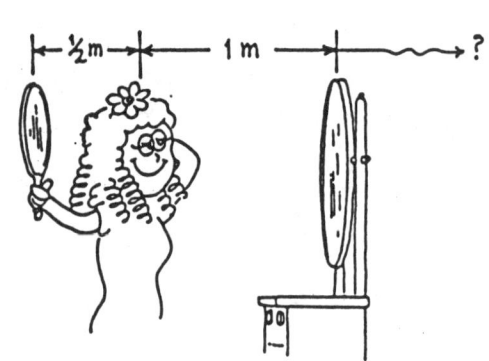

ANSWER:

2 meters in back of the dresser mirror. Why? Because the image of the flower in the small mirror is as far behind the small mirror as the flower is in front: ½ meter.

This puts the flower image a distance 1 + ½ + ½ meters in front of the dresser mirror. This image is just as far behind the dresser mirror --- 2 meters.

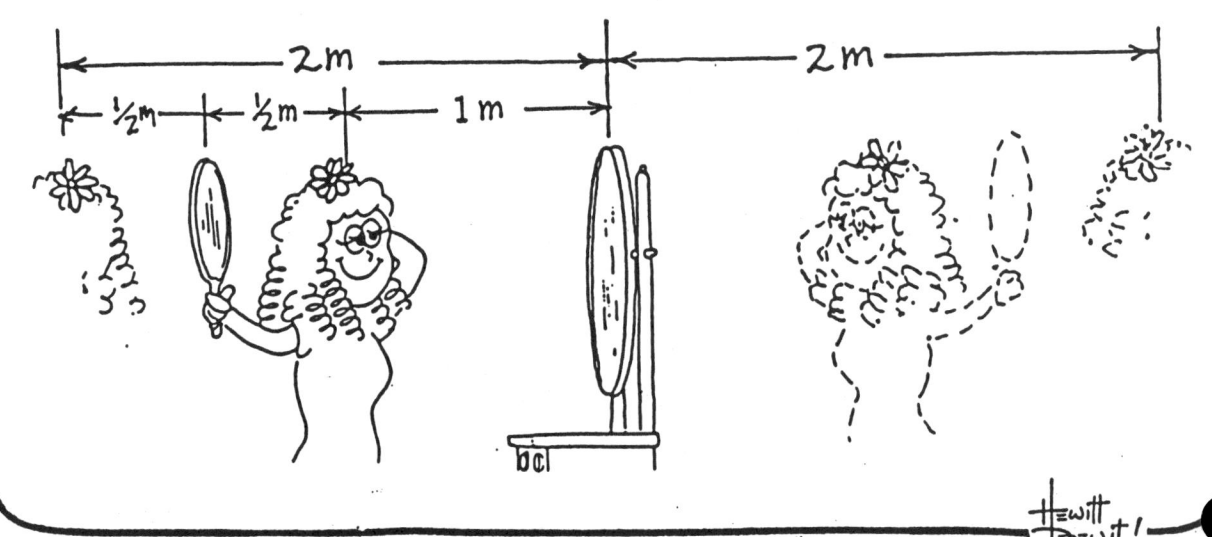

CHAP. 27

CONCEPTUAL Physics

She takes a photograph of her friend standing on the bridge as shown. Which of the two sketches more accurately shows the photograph of the bridge and its reflection?

CHAP. 27

CONCEPTUAL Physics

She takes a photograph of her friend standing on the bridge as shown. Which of the two sketches more accurately shows the photograph of the bridge and its reflection?

Answer:

The sketch on the right shows a more accurate reflection of the bridge. The reflected view is not simply an inversion of the scene above, as some people think, but is the scene as viewed from a lower position -- from the water surface. The reflected view of the bridge is the view the girl would see if her head were upside down at the water surface where the light is reflected. Hence the reflected view shows the underside of the bridge.

> Place a mirror flat on the floor between you and a table. Whereas the ordinary view shows the table top, the reflected view shows the bottom.

CHAP. 27

CONCEPTUAL Physics

Why does light from the sun or moon appear as a column when reflected from a body of water? How would it appear if the water surface were perfectly smooth?

CONCEPTUAL Physics

Why does light from the sun or moon appear as a column when reflected from a body of water? How would it appear if the water surface were perfectly smooth?

Answer:

If the water were perfectly smooth, a mirror image of the round sun or moon would be seen in the water. If the water were slightly rough, the image would be wavy. If the water were a bit more rough, little glimmers of the sun or moon would be seen above and below the main image. This is because the water waves act like an assemblage of small flat mirrors. For rougher waves, there is a greater variety of mirror facets properly tilted to reflect sunlight or moonlight into your eye. The light then appears smeared into a long vertical column.

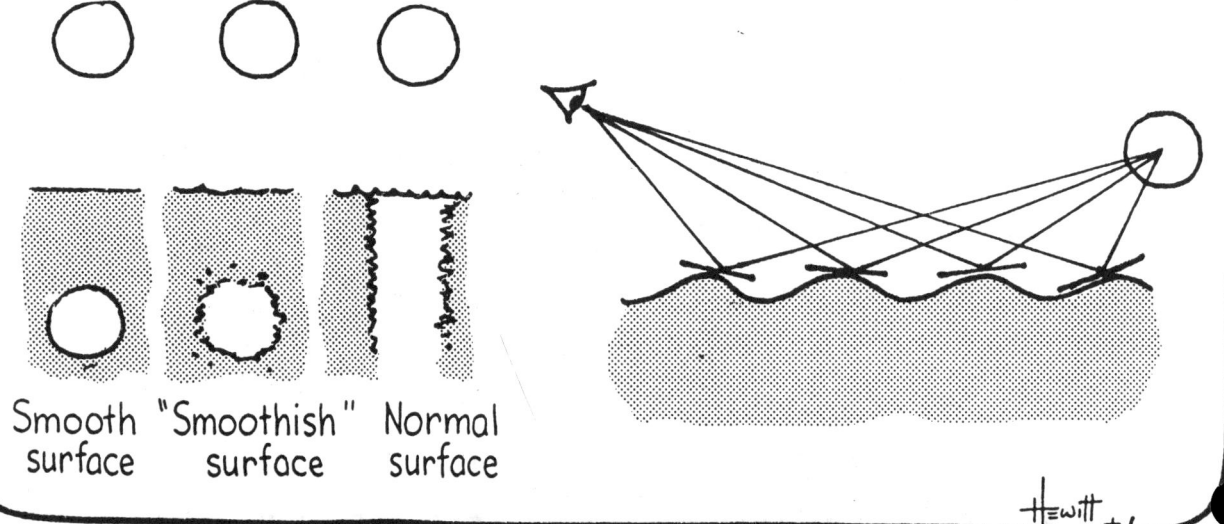

Smooth surface   "Smoothish" surface   Normal surface

CHAP. 27

CONCEPTUAL Physics

The photographer wishes to photograph the rainbow but is disappointed to find the camera's angle of view is not wide enough to see the whole rainbow. To get the whole rainbow, she would be better off if she were

a) closer to the rainbow

b) farther from the rainbow

c) ... neither, for she'd get the same portion of bow in either case

CONCEPTUAL Physics

The photographer wishes to photograph the rainbow but is disappointed to find the camera's angle of view is not wide enough to see the whole rainbow. To get the whole rainbow, she would be better off if she were

a) closer to the rainbow
b) farther from the rainbow
c) ... neither, for she'd get the same portion of bow in either case

The answer is C:

Any full circle rainbow, near or far, subtends an angle of 84°. So to photograph a full rainbow, whether a very close one produced by a hand-held garden hose or one miles away, the camera's field of view must be at least 84° — a very wide-angle lens. It is the angle of view, not the distance, that matters.

All rainbows, by the way, are completely round, as can be seen from a high-flying helicopter. Viewed from below, however, the ground gets in the way. To photograph a full-circle rainbow from a helicopter, the 84° field of view must be vertical as well as horizontal. Has anyone successfully taken a photograph of a full-circle rainbow?

CONCEPTUAL **Physics**

A COIN LIES SUBMERGED AT THE BOTTOM OF A PAN OF WATER. DOES REFRACTION OF LIGHT FROM THE COIN MAKE IT APPEAR DEEPER, OR MAKE IT APPEAR SHALLOWER THAN IT REALLY IS?

ANSWER:

THE COIN APPEARS SHALLOWER THAN IT REALLY IS. TO LOCATE THE IMAGE, DRAW AT LEAST TWO DIFFERENT RAYS FROM THE OBJECT AND NOTE WHERE THEY APPEAR TO MEET IF EXTENDED BACK-WARDS. THAT'S WHERE THE IMAGE IS SEEN. IF YOU LOOK STRAIGHT DOWN, THE IMAGE IS ONLY ¾ THE ACTUAL DEPTH.

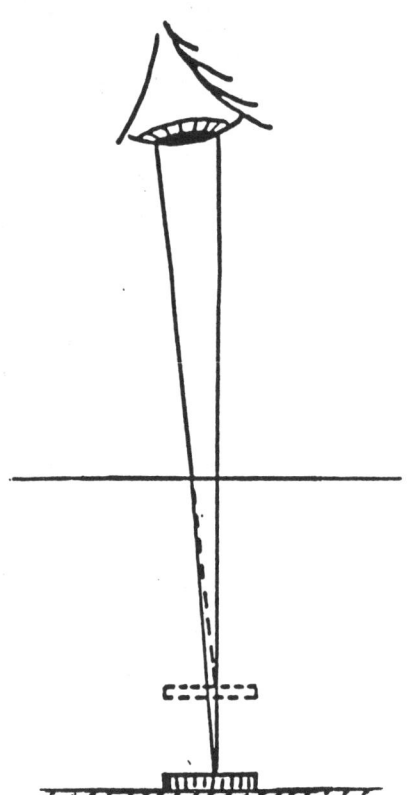

CONCEPTUAL Physics

Suppose you want to send a beam of laser light to a space station above the atmosphere and just above the horizon. You should aim your laser

a) slightly higher than

b) slightly lower than

c) directly along

the line of sight to the space station.

CHAP. 27

CONCEPTUAL **Physics**

Suppose you want to send a beam of laser light to a space station above the atmosphere and just above the horizon. You should aim your laser

a) slightly higher than

b) slightly lower than

c) directly along

the line of sight to the space station.

The answer is C:

To send light to the space station, make no corrections and simply aim at the station you see. All deviations due to atmospheric refraction in your line of sight will be the same for your laser beam -- principle of reciprocity.

How about if the laser is red and the space station blue?

CHAP. 27

CONCEPTUAL Physics

A PERSON WHO SEES MORE CLEARLY UNDER WATER THAN IN AIR WITHOUT EYEGLASSES IS
  a) NEARSIGHTED
  b) FARSIGHTED
  c) NEITHER

ANSWER:

NEARSIGHTED. THE SPEED OF LIGHT IN WATER IS LESS THAN IN AIR, SO THE CHANGE IN SPEED IS LESS AS LIGHT GOES FROM WATER TO YOUR EYE. LESS REFRACTION OCCURS. THIS MAKES ALL PEOPLE MORE FARSIGHTED UNDER WATER, WHICH IS ADVANTAGEOUS IF YOU'RE NEARSIGHTED. IF YOU'RE VERY NEARSIGHTED, THE IMAGE MAY FALL ON YOUR RETINA AND YOU'LL SEE AS CLEARLY UNDER WATER AS A PERSON WITH NORMAL VISION WHO WEARS AN AIR-ENCLOSED MASK.

WATER

NORMAL EYE UNDER WATER   NEARSIGHTED EYE UNDER WATER

ARE FISH NEARSIGHTED OR FARSIGHTED IN AIR?

CHAP. 27

CONCEPTUAL Physics

A LUNAR ECLIPSE OCCURS WHEN THE MOON PASSES INTO THE EARTH'S SHADOW. INSTEAD OF BEING COMPLETELY DARK, THE MOON APPEARS A DEEP RED. WHAT DOES THIS REDDISH COLOR HAVE TO DO WITH THE SUNSETS ALL OVER THE WORLD?

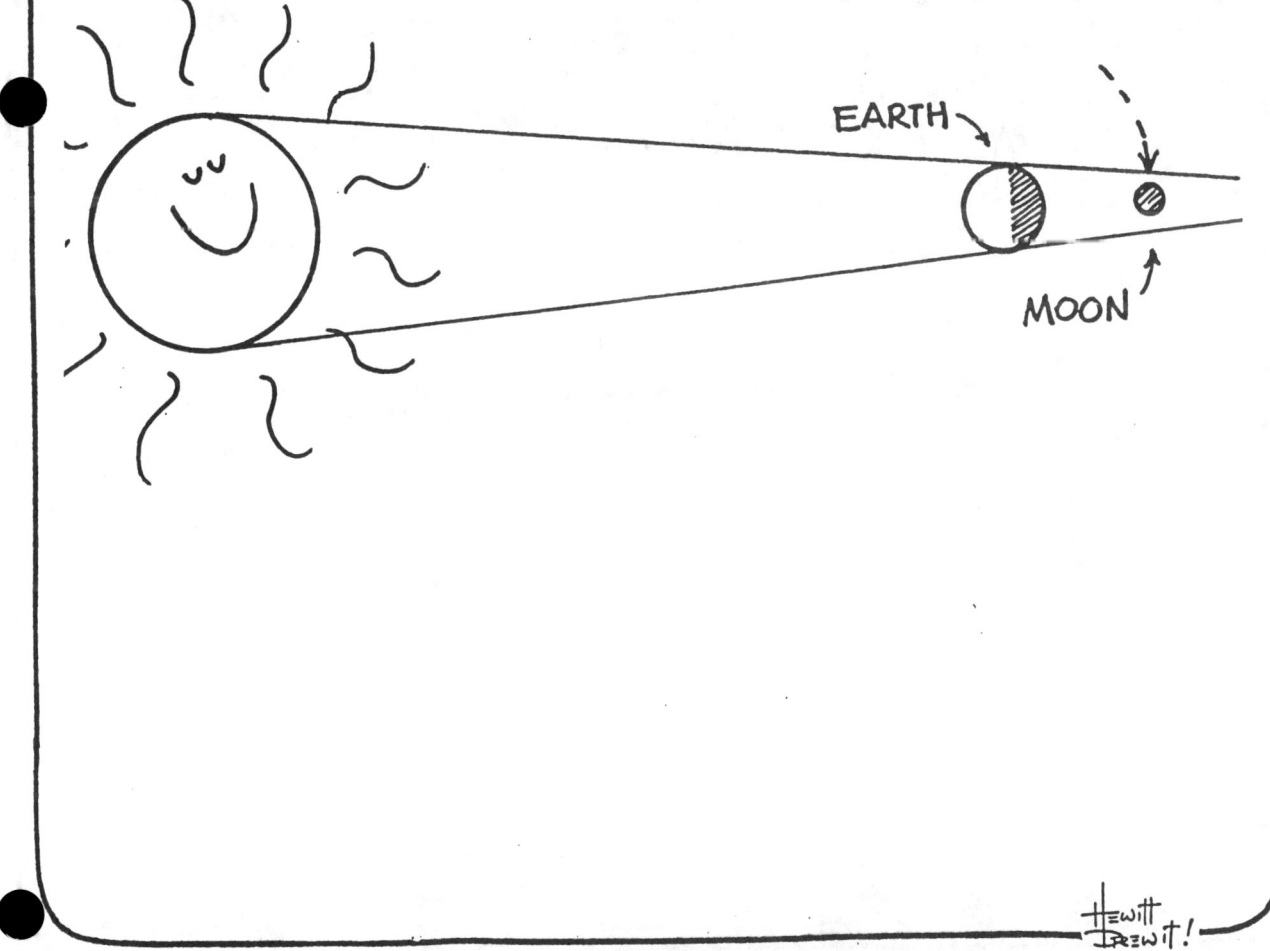

CONCEPTUAL Physics

A LUNAR ECLIPSE OCCURS WHEN THE MOON PASSES INTO THE EARTH'S SHADOW. INSTEAD OF BEING COMPLETELY DARK, THE MOON APPEARS A DEEP RED. WHAT DOES THIS REDDISH COLOR HAVE TO DO WITH THE SUNSETS ALL OVER THE WORLD?

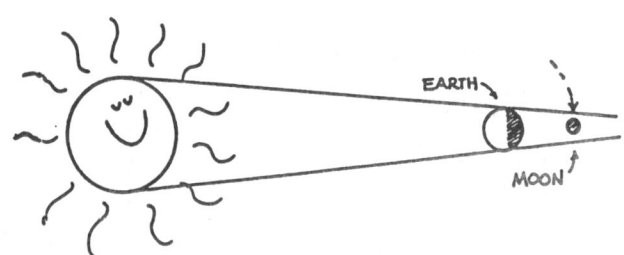

ANSWER:

DURING A LUNAR ECLIPSE, LIGHT FROM THE SUN GRAZES THE EARTH'S ATMOSPHERE WHICH ACTS LIKE A LENS TO REFRACT LIGHT ONTO THE OTHERWISE DARK MOON. ONLY THE LOW FREQUENCIES TRAVERSE THE LONG PATH THROUGH THE ATMOSPHERE. SO THE LIGHT TO FALL UPON THE MOON IS THE RED AND ORANGE LIGHT REFRACTED BY ALL THE SUNSETS, A FULL 360°, ALL AROUND THE WORLD.

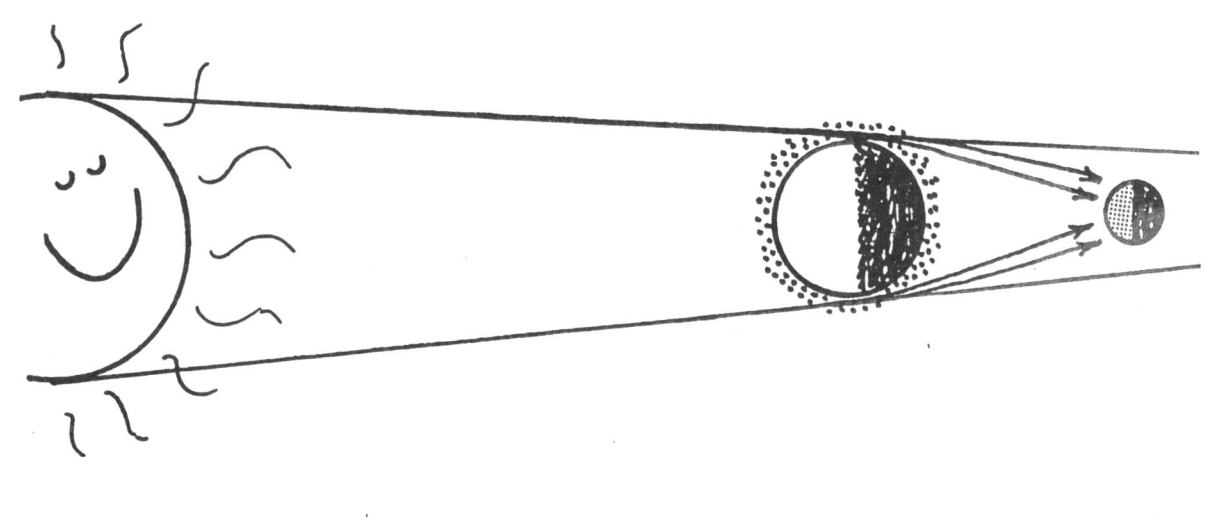

CONCEPTUAL Physics

THE LINES ON THE LENSES OF THE EYEGLASSES INDICATE THE PLANE OF POLARIZATION THROUGH WHICH LIGHT CAN PASS. WHICH PAIR OF GLASSES HAS THE PLANE OF POLARIZATION IN THE BEST ORIENTATION FOR REDUCING ROAD GLARE WHILE DRIVING?

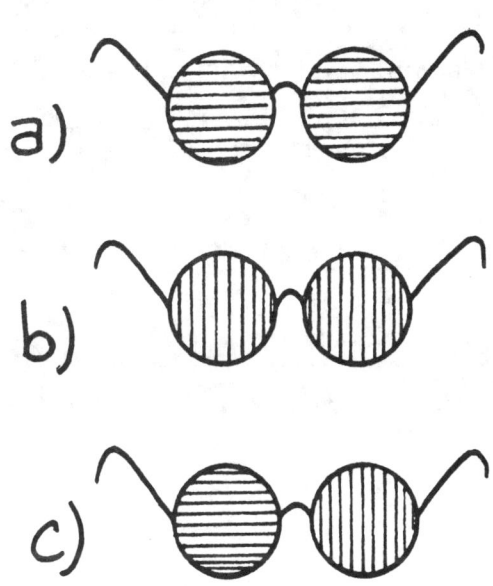

CHAP. 28

CONCEPTUAL Physics

THE LINES ON THE LENSES OF THE EYEGLASSES INDICATE THE PLANE OF POLARIZATION THROUGH WHICH LIGHT CAN PASS. WHICH PAIR OF GLASSES HAS THE PLANE OF POLARIZATION IN THE BEST ORIENTATION FOR REDUCING ROAD GLARE WHILE DRIVING?

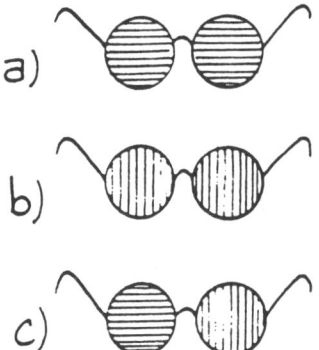

ANSWER:

GLASSES **b** ARE BEST FOR REDUCING ROAD GLARE BECAUSE MOST OF THE LIGHT THAT COMPOSES GLARE FROM NONMETALLIC SURFACES IS POLARIZED IN THE SAME PLANE AS THE SURFACE -- HORIZONTAL. GLASSES **a** WILL PASS HORIZONTALLY POLARIZED LIGHT AND WILL BE TERRIBLE FOR DRIVING. GLASSES **C** ARE FOR THE PERSON WHO WANTS ALL BETS COVERED. WHEN THE GLARE IS FROM HORIZONTAL SURFACES, THE LEFT EYE SHOULD BE CLOSED; WHEN THE GLARE IS FROM VERTICAL SURFACES, FLAGPOLES AND THE LIKE, THE RIGHT EYE SHOULD BE CLOSED. SINCE MOST GLARE IS FROM HORIZONTAL SURFACES, AND DRIVING IS BEST WITH BOTH EYES, SELECT **b**!

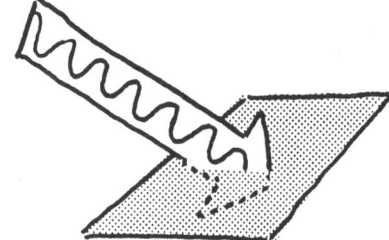

CHAP. 28

CONCEPTUAL Physics

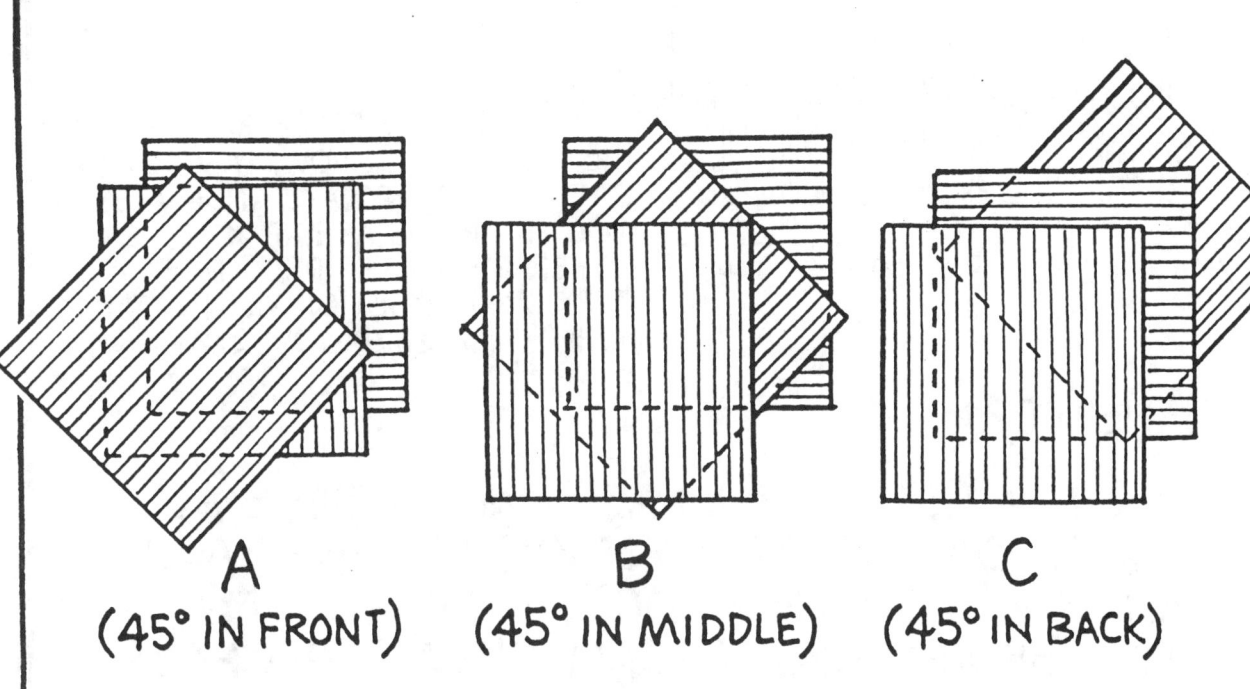

A
(45° IN FRONT)

B
(45° IN MIDDLE)

C
(45° IN BACK)

Three sets of Polaroids, one atop the other, are shown above. In each set the polarization axes of two polaroids are at 90° to each other, and a third is at 45° to the two. Which set(s) will pass light where the three overlap?

CHAP. 28

CONCEPTUAL Physics

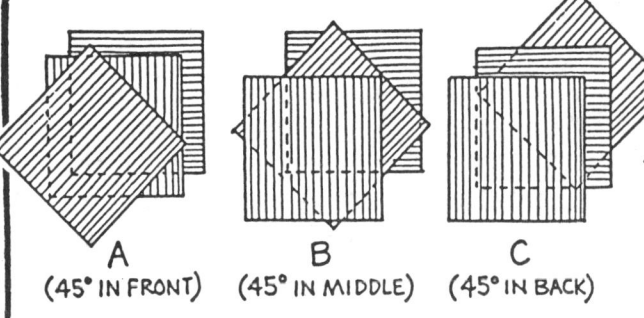

A (45° IN FRONT)   B (45° IN MIDDLE)   C (45° IN BACK)

THREE SETS OF POLAROIDS, ONE ATOP THE OTHER, ARE SHOWN ABOVE. IN EACH SET THE POLARIZATION AXES OF TWO POLAROIDS ARE AT 90° TO EACH OTHER, AND A THIRD IS AT 45° TO THE TWO. WHICH SET(S) WILL PASS LIGHT WHERE THE THREE OVERLAP?

ANSWER:

ONLY SET B WILL PASS LIGHT WHERE ALL POLAROIDS OVERLAP, FOR THE AXIS OF EACH POLAROID IS NOT AT 90° TO THE ONE NEXT TO IT. THE VECTOR DIAGRAM SHOWS THAT HALF THE LIGHT GETS THROUGH THE FIRST POLAROID, SHOWN BY THE VERTICAL VECTOR, AND 0.707 OF THIS GETS THROUGH THE SECOND POLAROID BECAUSE IT IS AT 45° (NOT 90°!), AND IN TURN 0.707 OF THIS GETS THROUGH THE THIRD.

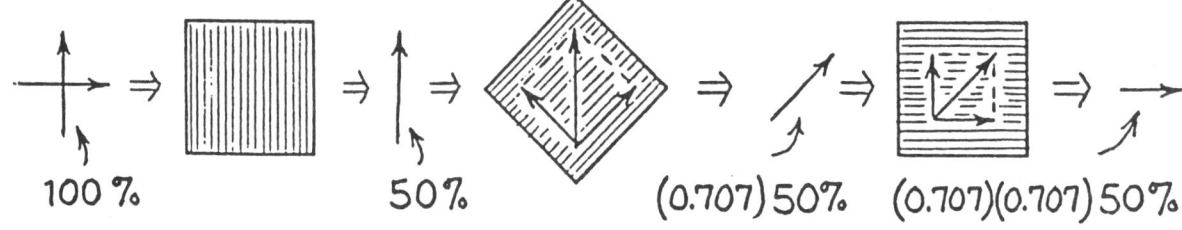

100%   50%   (0.707) 50%   (0.707)(0.707) 50%

IN SET A ALL LIGHT IS BLOCKED BY THE BACK PAIR OF POLAROIDS BECAUSE THEY ARE AT 90° TO EACH OTHER. LIKEWISE WITH THE FRONT PAIR OF POLAROIDS IN SET C.

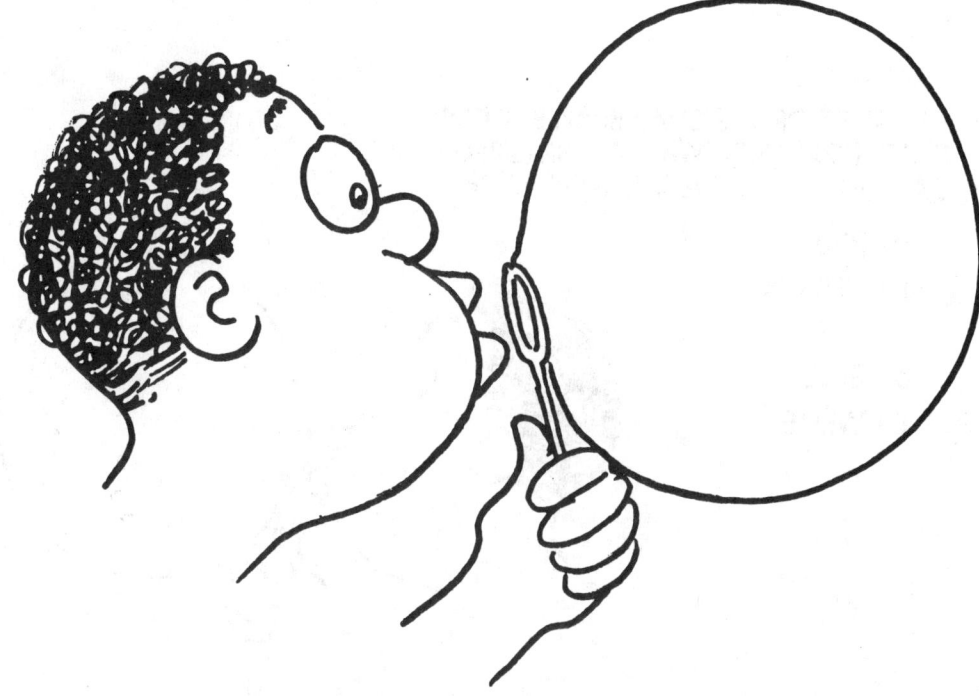

CONCEPTUAL Physics

Part of the soap bubble looks blue in sunlight. What color is being cancelled by wave interference?

a) RED
b) YELLOW
c) GREEN
d) BLUE
e) WHITE

CONCEPTUAL Physics

PART OF THE SOAP BUBBLE LOOKS BLUE IN SUNLIGHT. WHAT COLOR IS BEING CANCELLED BY WAVE INTERFERENCE?

a) RED
b) YELLOW
c) GREEN
d) BLUE
e) WHITE

THE ANSWER IS b:

THE PART OF THE BUBBLE THAT LOOKS BLUE IS DEFICIENT IN ITS COMPLEMENTARY COLOR, YELLOW.

CONCEPTUAL Physics

WHICH OF THESE LAMPS IS EMITTING ELECTROMAGNETIC RADIATION?

a) LAMP A
b) LAMP B
c) BOTH
d) NEITHER

CONCEPTUAL Physics

WHICH OF THESE LAMPS IS EMITTING ELECTROMAGNETIC RADIATION?

a) LAMP A
b) LAMP B
c) BOTH
d) NEITHER

THE ANSWER IS **C**, BOTH:

ALL BODIES WITH ANY TEMPERATURE AT ALL ARE CONTINUALLY EMITTING ELECTROMAGNETIC WAVES. THE FREQUENCY OF THESE WAVES VARIES WITH TEMPERATURE, $\bar{f} \sim T$. LAMP B IS HOT ENOUGH SO THAT THE WAVES IT EMITS ARE VISIBLE LIGHT. LAMP A IS COOLER, AND THE RADIATION IT EMITS IS TOO LOW IN FREQUENCY TO BE VISIBLE -- IT EMITS INFRARED WAVES, WHICH AREN'T SEEN WITH THE EYE. YOU EMIT WAVES AS WELL. EVEN IN A COMPLETELY DARK ROOM YOUR WAVES ARE THERE. YOUR FRIENDS MAY NOT BE ABLE TO SEE YOU, BUT A RATTLESNAKE CAN!

CONCEPTUAL Physics

WHEN THE ZINC BALL ON THE CHARGED ELECTROSCOPE IS ILLUMINATED WITH ULTRAVIOLET LIGHT, THE LEAVES OF THE ELECTROSCOPE COLLAPSE. WAS THE ELECTROSCOPE CHARGED POSITIVE, OR NEGATIVE?

CONCEPTUAL Physics

When the zinc ball on the charged electroscope is illuminated with ultraviolet light, the leaves of the electroscope collapse. Was the electroscope charged positive, or negative?

ANSWER:

The electroscope was discharged by the photoelectric effect. UV light incident upon the zinc dislodged electrons into the air. Hence the electroscope must have been negatively charged. If it were positively charged, the dislodging of electrons would have made it more positively charged, not less charged. Incidentally, if glass were placed between the UV source and the zinc, no discharge would occur. Why? Because glass acts as a filter for UV light.

CHAP. 30

CONCEPTUAL Physics

COMPARED TO HYDROGEN, ^1_1H, THE ELEMENT HELIUM, ^4_2He HAS

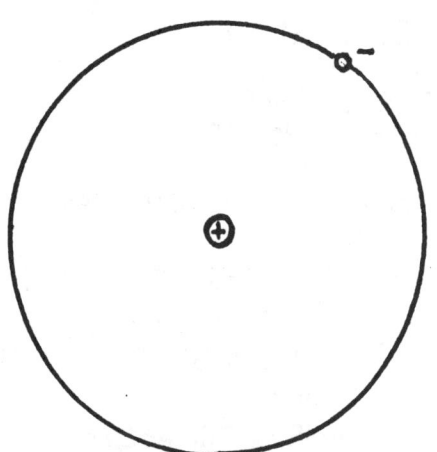

a) MORE MASS AND IS LARGER IN SIZE

b) MORE MASS AND IS ABOUT THE SAME IN SIZE

c) MORE MASS AND IS SMALLER IN SIZE

d) NONE OF THE ABOVE

CONCEPTUAL Physics

COMPARED TO HYDROGEN, ^1_1H, THE ELEMENT HELIUM, ^4_2He HAS

a) MORE MASS AND IS LARGER IN SIZE
b) MORE MASS AND IS ABOUT THE SAME IN SIZE
c) MORE MASS AND IS SMALLER IN SIZE
d) NONE OF THE ABOVE

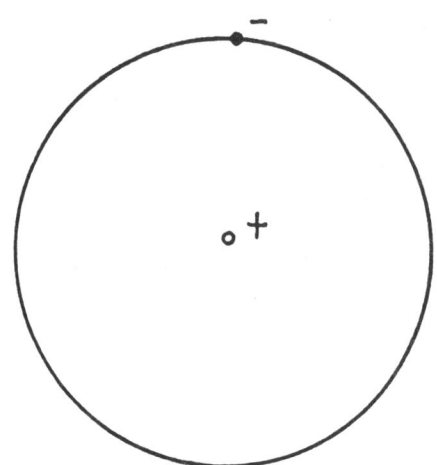

THE ANSWER IS C:

THE NUCLEUS OF HELIUM HAS FOUR NUCLEONS COMPARED TO HYDROGEN'S ONE, SO IT IS ABOUT FOUR TIMES AS MASSIVE AS HYDROGEN. THE NUCLEUS OF HELIUM HAS TWICE THE ELECTRIC CHARGE OF HYDROGEN, AND PULLS ITS ELECTRONS INTO A TIGHTER ORBIT THAN HYDROGEN. HELIUM IS A SMALLER BUT HEAVIER ATOM THAN HYDROGEN.

CONCEPTUAL Physics

An archeologist extracts a sample of carbon from an ancient ax handle and finds that it emits an average of 10 beta emissions per minute. She finds that the same mass of carbon from a living tree emits 40 betas per minute.

Knowing that the half life of carbon-14 is 5730 years, she concludes that the age of the ax handle is about

a) 2865 years
b) 5730 years
c) 11460 years
d) 17190 years

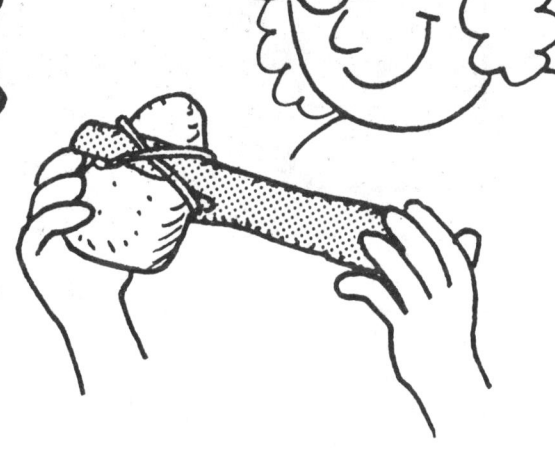

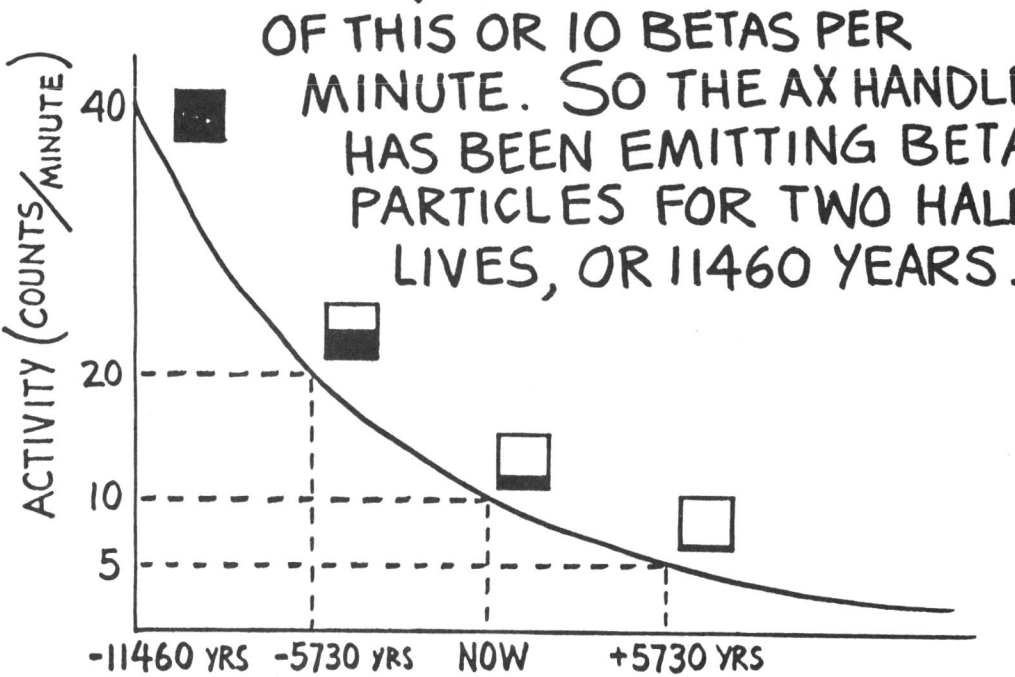

CONCEPTUAL Physics

Suppose you are given three radioactive cookies — one an alpha emitter, one a beta emitter, one a gamma emitter. You must eat one, hold one in your hand, and put the other in your pocket. What can you do to minimize your exposure to radiation?

CHAP. 32

CONCEPTUAL Physics

Suppose you are given three radioactive cookies — one an alpha emitter, one a beta emitter, one a gamma emitter. You must eat one, hold one in your hand, and put the other in your pocket. What can you do to minimize your exposure to radiation?

Answer:

Ideally get as far from the cookies as possible. But if you must eat one, hold one, and put one in your pocket, hold the alpha; the skin on your hand will shield you. Put the beta in your pocket; your clothing will likely shield you. Eat the gamma; it will penetrate your body in any of these cases anyway.

CONCEPTUAL Physics

IF YOU BEGIN WITH 120 GRAMS OF PURE RADIOACTIVE CHROMIUM-55 WHICH UNDERGOES BETA DECAY WITH A HALF LIFE OF 3.5 MINUTES, THEN AFTER 14 MINUTES YOUR SAMPLE WILL BE PRIMARILY

a) CHROMIUM
b) VANADIUM
c) IRON
d) TITANIUM
e) MANGANESE

AND THE AMOUNT OF CHROMIUM-55 LEFT WILL BE

f) 60 GRAMS
g) 30 GRAMS
h) 15 GRAMS
i) 7.5 GRAMS

CHAP. 32

CONCEPTUAL Physics

> IF YOU BEGIN WITH 120 GRAMS OF PURE RADIOACTIVE CHROMIUM-55 WHICH UNDERGOES BETA DECAY WITH A HALF LIFE OF 3.5 MINUTES, THEN AFTER 14 MINUTES YOUR SAMPLE WILL BE PRIMARILY
>
> a) CHROMIUM
> b) VANADIUM
> c) IRON
> d) TITANIUM
> e) MANGANESE

> AND THE AMOUNT OF CHROMIUM-55 LEFT WILL BE
>
> f) 60 GRAMS
> g) 30 GRAMS
> h) 15 GRAMS
> i) 7.5 GRAMS

THE ANSWERS ARE e AND i:

BETA EMISSION IS THE EMISSION OF AN ELECTRON, WHICH EFFECTIVELY TRANSFORMS A NEUTRON TO A PROTON, INCREASING THE ATOMIC NUMBER OF THE RADIOACTIVE ELEMENT BY 1. A LOOK AT THE PERIODIC TABLE SHOWS THAT MANGANESE, ATOMIC NUMBER 25, IS 1 GREATER THAN CHROMIUM, ATOMIC NUMBER 24. AFTER 14 MINUTES, 4 HALF LIVES OF 3.5 MINUTES EACH HAVE ELAPSED (14/3.5 = 4). THE 120 GRAMS IS THEREFORE HALVED 4 TIMES TO LEAVE 7.5 GRAMS. THE REMAINING MATERIAL IS MANGANESE.

CHAP. 32

When a neutron interacts with a U-235 nucleus, it can fission many possible ways. What element results if it fissions into two identical nuclei?

Can you answer this one? How many neutrons are produced when a U-235 nucleus fissions into Sr-90 and Xe-138?

CONCEPTUAL Physics

> WHEN A NEUTRON INTERACTS WITH A U-235 NUCLEUS, IT CAN FISSION MANY POSSIBLE WAYS. WHAT ELEMENT RESULTS IF IT FISSIONS INTO TWO IDENTICAL NUCLEI?

> CAN YOU ANSWER THIS ONE? HOW MANY NEUTRONS ARE PRODUCED WHEN A U-235 NUCLEUS FISSIONS INTO Sr-90 AND Xe-138?

ANSWERS:

IF URANIUM FISSIONS INTO TWO IDENTICAL ELEMENTS, THEIR ATOMIC NUMBER IS HALF 92, OR 46. THAT'S PALLADIUM.

IF U-235 FISSIONS INTO STRONTIUM-90 AND XENON-138, 8 NEUTRONS ARE RELEASED, ACCORDING TO THE REACTION

$$^{235}_{92}U + ^{1}_{0}n \longrightarrow ^{90}_{38}Sr + ^{138}_{54}Xe + 8(^{1}_{0}n).$$

(THE NUMBER OF NEUTRONS RELEASED PER FISSION REACTION FOR MOST REACTIONS IS CONSIDERABLY LESS THAN 8.)

CHAP. 32

CONCEPTUAL Physics

WHAT DOES THE RADIOACTIVE DECAY OF URANIUM TO LEAD HAVE TO DO WITH THE GAS IN THE CHILD'S BALLOON?

ANSWER:

THE GAS IN THE BALLOON IS HELIUM. THESE HELIUM ATOMS WERE ONCE THE ALPHA PARTICLES OF RADIOACTIVE DECAY TRAPPED WITH OTHER PARTICLES BENEATH THE GROUND IN NATURAL GAS DEPOSITS. FOR EACH URANIUM ATOM THAT DECAYS TO LEAD, 8 ALPHA PARTICLES ARE EMITTED. AN ALPHA PARTICLE WITH 2 ELECTRONS IS A HELIUM ATOM.

LIKE A CANNONBALL, AN ALPHA PARTICLE IS HARMFUL ONLY WHEN IT HAS A HIGH KINETIC ENERGY.

CHAP. 32

CONCEPTUAL Physics

She bathes in the warmth of a natural hot spring located in the quiet and peaceful mountains. Interestingly enough, the spring water is warmed by

- a) FIRES BENEATH THE EARTH'S SURFACE
- b) THE EARTH'S OWN NATURAL HEAT
- c) SOLAR POWER
- d) NUCLEAR POWER

ANSWER:

The spring is warmed by nuclear power -- not from nuclear power plants that generate electricity -- but from the natural radioactive decay of atomic nuclei in the earth's interior. Radioactivity is the source of the earth's "own natural heat," which produces hot springs -- and geysers as well.

> This doesn't mean that hot springs and geysers themselves are radioactive. Their thermal energy is simply a byproduct of nuclear decay deep beneath the earth's surface.

CHAP. 32

CONCEPTUAL Physics

Is the decay $^{16}_{8}O \rightarrow {}^{12}_{6}C + {}^{4}_{2}He$ possible? And if so, would this reaction require energy, or go by itself and yield energy?

ANSWER:

THE DECAY IS POSSIBLE, FOR BOTH CHARGE AND MASS NUMBER ARE CONSERVED (THE NUMBERS ON THE LEFT = THE TOTAL NUMBERS ON THE RIGHT). INSPECTION OF THE MASS/NUCLEON-VS-ATOMIC-NUMBER CURVE SHOW THAT THE REACTION "CLIMBS" THE HILL, SO THERE IS MORE MASS AFTER THE REACTION THAN BEFORE. THIS MEANS THE REACTION WILL REQUIRE AN INPUT OF ENERGY. HOW MUCH? AN AMOUNT EQUAL TO THE GAIN IN MASS MULTIPLIED BY THE SPEED OF LIGHT SQUARED!

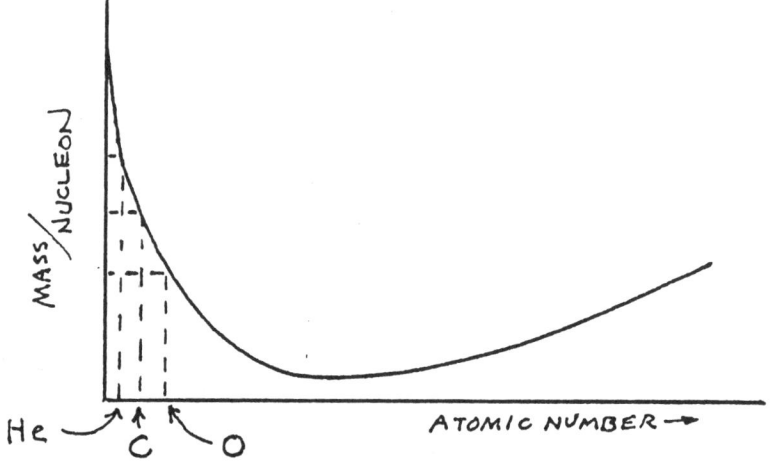

CONCEPTUAL Physics

Some common fusion reactions of hydrogen isotopes are shown in incomplete form below. Can you complete them?

$$^{2}_{1}H + ^{2}_{1}H \rightarrow ^{3}_{2}He + (\quad)$$

$$^{2}_{1}H + ^{3}_{1}H \rightarrow ^{4}_{2}He + (\quad)$$

$$^{3}_{1}H + (\quad) \rightarrow ^{4}_{2}He + ^{1}_{0}n + ^{1}_{0}n$$

CONCEPTUAL Physics

Some common fusion reactions of hydrogen isotopes are shown in incomplete form below. Can you complete them?

$^2_1H + {}^2_1H \rightarrow {}^3_2He + (\quad)$

$^2_1H + {}^3_1H \rightarrow {}^4_2He + (\quad)$

$^3_1H + (\quad) \rightarrow {}^4_2He + {}^1_0n + {}^1_0n$

ANSWERS:

$^2_1H + {}^2_1H \rightarrow {}^3_2He + {}^1_0n$

$^2_1H + {}^3_1H \rightarrow {}^4_2He + {}^1_0n$

$^3_1H + {}^3_1H \rightarrow {}^4_2He + {}^1_0n + {}^1_0n$

CHAP. 33

CONCEPTUAL Physics

An astronaut ages 3 years when traveling at 99% the speed of light to the star Procyon and back. The space officials to greet her on her return age

a) Less than 3 years
b) 3 years
c) More than 3 years

CHAP. 34

CONCEPTUAL Physics

An astronaut ages 3 years when traveling at 99% the speed of light to the star Procyon and back. The space officials to greet her on her return age
  a) less than 3 years
  b) 3 years
  c) more than 3 years

THE ANSWER IS C:

The 3 years experienced by the traveling astronaut is considerably less than if she had stayed at home. The stay-at-homes age more than 3 years --- 21.2 years to be exact.

$$t = \frac{t_o}{\sqrt{1-(\frac{v}{c})^2}} = \frac{3 \text{ YRS}}{\sqrt{1-(\frac{0.99c}{c})^2}} = \frac{3 \text{ YRS}}{\sqrt{1-0.99^2}} = \frac{3 \text{ YRS}}{\sqrt{0.02}} = 21.2 \text{ YRS}$$

CHAP. 34

CONCEPTUAL Physics

A 1-METER LONG SPEAR IS THROWN AT A RELATIVISTIC SPEED THROUGH A PIPE THAT IS 1 METER LONG. BOTH THESE DIMENSIONS ARE MEASURED WHEN EACH IS AT REST. WHEN THE SPEAR PASSES THROUGH THE PIPE, WHICH OF THESE STATEMENTS BEST DESCRIBES WHAT IS OBSERVED?

a) THE SPEAR SHRINKS SO THAT THE PIPE COMPLETELY COVERS IT AT SOME POINT

b) THE PIPE SHRINKS SO THAT THE SPEAR EXTENDS FROM BOTH ENDS AT SOME POINT

c) BOTH SHRINK EQUALLY SO THE PIPE COMPLETELY COVERS IT AT SOME POINT

d) ANY OF THESE, DEPENDING ON THE MOTION OF THE OBSERVER

CHAP. 34

CONCEPTUAL Physics

A 1-METER LONG SPEAR IS THROWN AT A RELATIVISTIC SPEED THROUGH A PIPE THAT IS 1 METER LONG. BOTH THESE DIMENSIONS ARE MEASURED WHEN EACH IS AT REST. WHEN THE SPEAR PASSES THROUGH THE PIPE, WHICH OF THESE STATEMENTS BEST DESCRIBES WHAT IS OBSERVED?

a) THE SPEAR SHRINKS SO THAT THE PIPE COMPLETELY COVERS IT AT SOME POINT

b) THE PIPE SHRINKS SO THAT THE SPEAR EXTENDS FROM BOTH ENDS AT SOME POINT

c) BOTH SHRINK EQUALLY SO THE PIPE COMPLETELY COVERS IT AT SOME POINT

d) ANY OF THESE, DEPENDING ON THE MOTION OF THE OBSERVER

THE ANSWER IS d:

OBSERVE FROM A REST POSITION WITH RESPECT TO THE PIPE AND AT SOME POINT THE CONTRACTED SPEAR WILL BE COMPLETELY COVERED BY THE PIPE. OR TRAVEL ALONG WITH THE SPEAR AND YOU'LL SEE THE SPEAR AT SOME POINT EXTEND FROM THE CONTRACTED PIPE. OR MOVE BETWEEN THE SPEAR AND THE PIPE AT A CERTAIN INTERMEDIATE VELOCITY, AND SEE BOTH THE SPEAR AND THE PIPE CONTRACTED THE SAME AMOUNT. SO WHAT REALLY HAPPENS IS RELATIVE --- IT DEPENDS ON YOUR POINT OF VIEW, OR FRAME OF REFERENCE!

CHAP. 34

CONCEPTUAL Physics

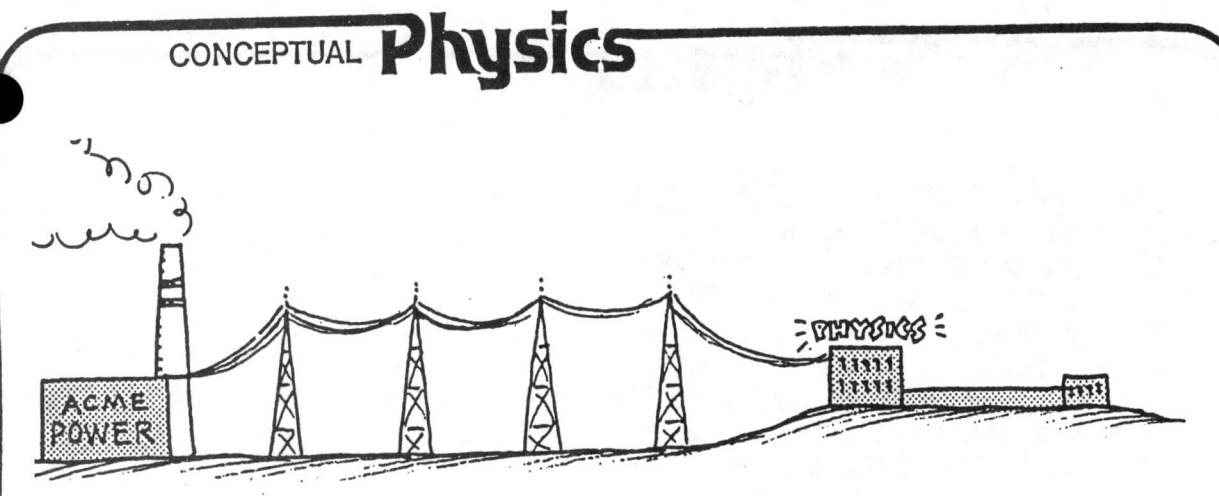

In a particle accelerator, the masses of particles accelerated to nearly the speed of light increase by more than 1000 times. The power utility that supplies energy to the accelerator obtains its energy from fuels, which decrease in mass by either combustion or fission. How does the increase in the masses of the accelerated particles compare to the mass decrease at the power plant?

CONCEPTUAL Physics

IN A PARTICLE ACCELERATOR, THE MASSES OF PARTICLES ACCELERATED TO NEARLY THE SPEED OF LIGHT INCREASE BY MORE THAN 1000 TIMES. THE POWER UTILITY THAT SUPPLIES ENERGY TO THE ACCELERATOR OBTAINS ITS ENERGY FROM FUELS, WHICH DECREASE IN MASS BY EITHER COMBUSTION OR FISSION. HOW DOES THE INCREASE IN THE MASSES OF THE ACCELERATED PARTICLES COMPARE TO THE MASS DECREASE AT THE POWER PLANT?

ANSWER:

TAKING INEFFICIENCIES INTO ACCOUNT, GRAMS OF FUEL END UP AS MILLIGRAMS OF ACCELERATED PARTICLES. BUT IF WE NEGLECT THE INEFFICIENCIES OF THE POWER PLANT, AND MAINLY OF THE ACCELERATOR, THE MANY FUEL FRAGMENTS LOSE JUST AS MUCH TOTAL MASS AS THE RELATIVELY FEWER NUMBER OF ACCELERATED PARTICLES GAIN! TO SAY THAT A POWER UTILITY DELIVERS ENERGY IS TO SAY IT DELIVERS MASS, BECAUSE MASS AND ENERGY ARE ONE AND THE SAME.

CONCEPTUAL Physics

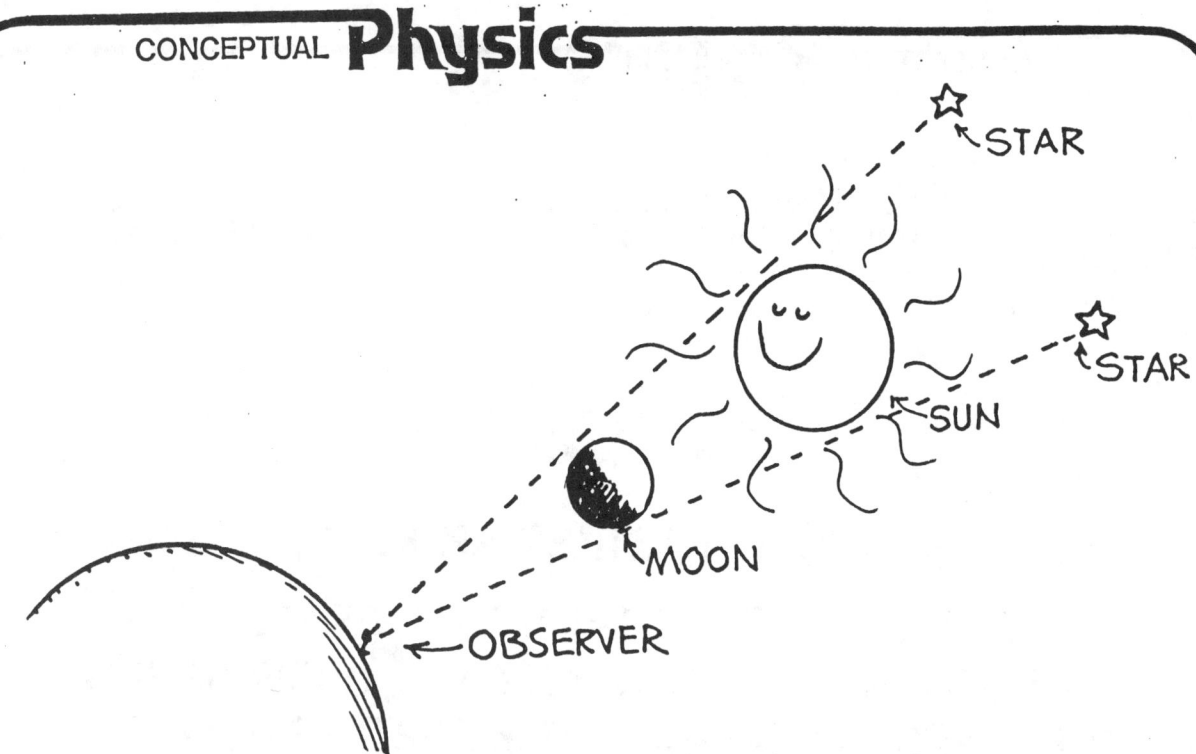

IF THE SUN PASSES BETWEEN THE EARTH AND A PAIR OF STARS AS SHOWN, AND THE MOON PASSES IN FRONT OF THE SUN AND TOTALLY ECLIPSES IT SO THE STARS ARE VISIBLE, THEN ACCORDING TO GENERAL RELATIVITY, THE STARS WILL APPEAR TO BE SLIGHTLY

- a) CLOSER TOGETHER
- b) FARTHER APART
- c) DISTORTED, BUT NOT CLOSER OR FARTHER APART

CONCEPTUAL Physics

IF THE SUN PASSES BETWEEN THE EARTH AND A PAIR OF STARS AS SHOWN, AND THE MOON PASSES IN FRONT OF THE SUN AND TOTALLY ECLIPSES IT SO THE STARS ARE VISIBLE, THEN ACCORDING TO GENERAL RELATIVITY, THE STARS WILL APPEAR TO BE SLIGHTLY

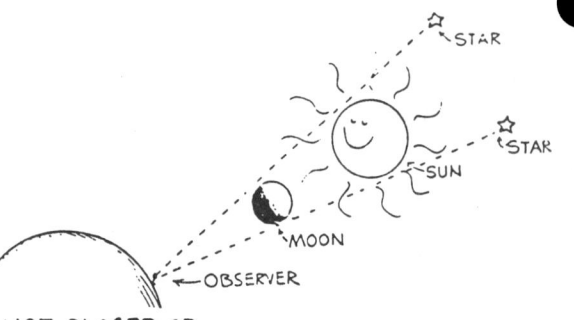

a) CLOSER TOGETHER
b) FARTHER APART
c) DISTORTED, BUT NOT CLOSER OR FARTHER APART

THE ANSWER IS b, FARTHER APART:

LIGHT FROM THE STARS THAT GRAZES THE SUN BENDS AS SHOWN BELOW. CONSEQUENTLY, THE STARS APPEAR SLIGHTLY FARTHER APART. THIS WAS PREDICTED BY EINSTEIN IN 1916 AND TESTED DURING THE TOTAL ECLIPSE OF THE SUN IN 1919. IT WAS THE FIRST CONFIRMATION OF EINSTEIN'S GENERAL THEORY OF RELATIVITY.

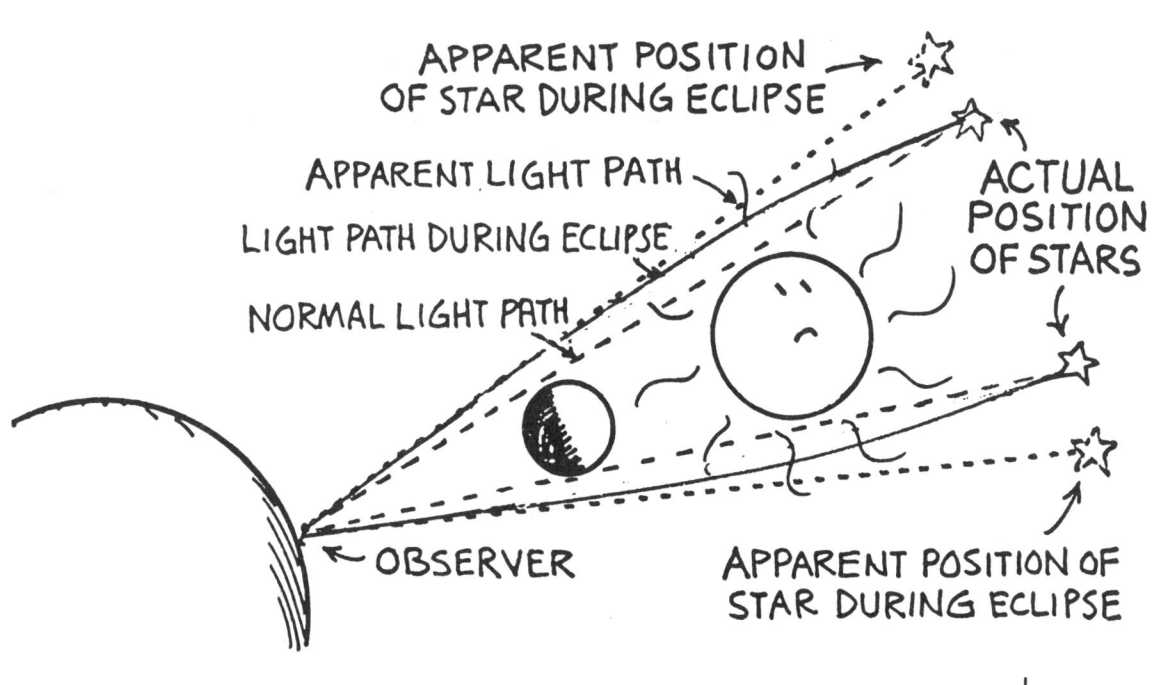

CONCEPTUAL Physics

Sailing is fun, especially on a windy day. Consider the top views of the two boats below, one sailing with the wind, and the other across the wind. Which can sail faster than wind speed?

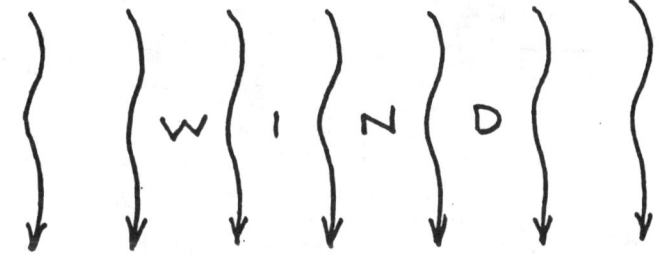

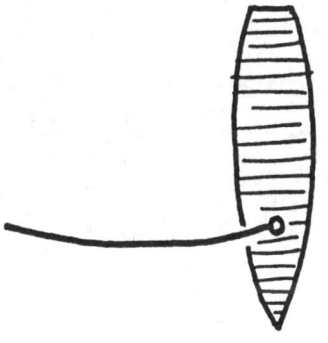
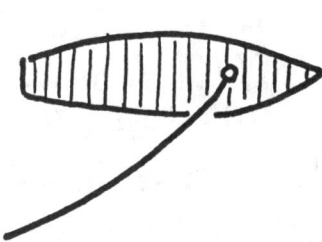

APPENDIX III

CONCEPTUAL Physics

Sailing is fun, especially on a windy day. Consider the top views of the two boats below, one sailing with the wind, and the other across the wind. Which can sail faster than wind speed?

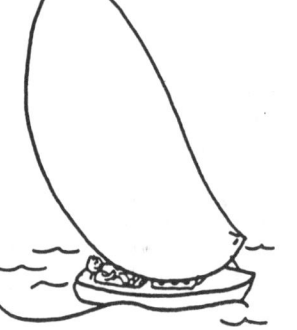

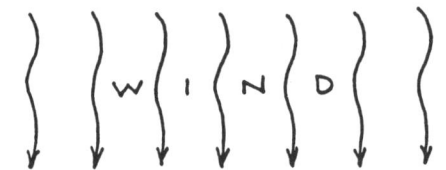

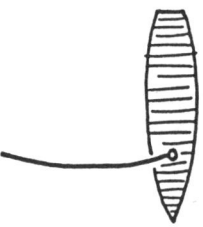 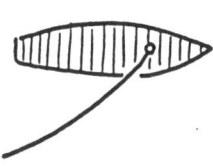

ANSWER:

The boat that sails directly with the wind can sail no faster than wind speed. Why? Even sailing as fast as the wind, there would be no wind impact against the sail. It would sag. But when sailing crosswind, there would still be wind impact against the sail, and speeds greater than wind speed can be achieved.

(Why will the sail also sag when at an angle of 45° and the boat travels crosswind at wind speed?)

CONCEPTUAL Physics

WE ALL KNOW THAT THE FORCE OF WIND IMPACT DRIVES A SAILBOAT. IN WHICH OF THE THREE POSITIONS DOES THE WIND IMPACT FORCE ACTUALLY INCREASE AS THE BOAT MOVES FASTER?

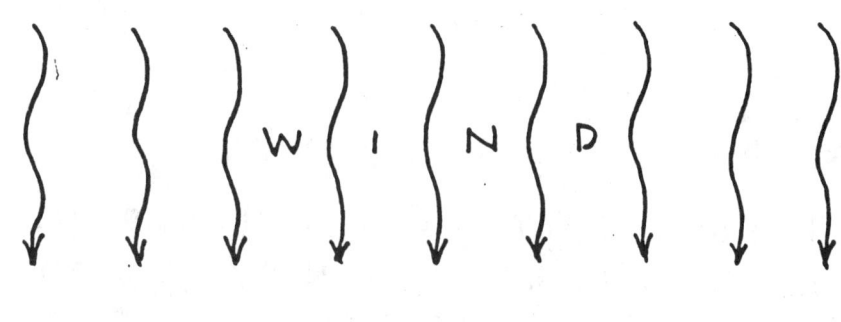

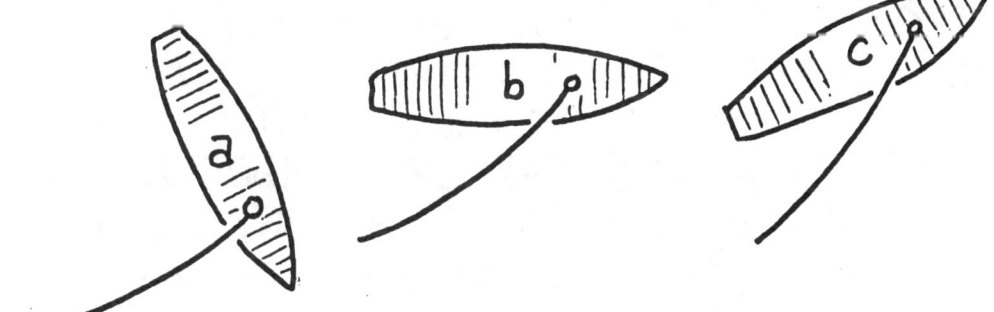

APPENDIX III

CONCEPTUAL Physics

WE ALL KNOW THAT THE FORCE OF WIND IMPACT DRIVES A SAILBOAT. IN WHICH OF THE THREE POSITIONS DOES THE WIND IMPACT FORCE ACTUALLY INCREASE AS THE BOAT MOVES FASTER?

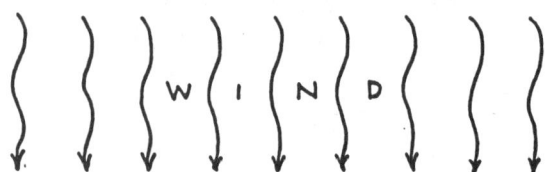

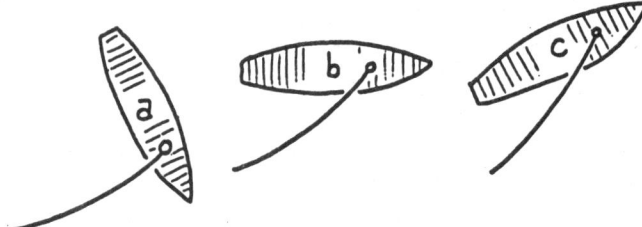

ANSWER:

JUST AS RAINDROPS HIT YOU HARDER THE FASTER YOU RUN INTO A SLANTING RAIN, THE WIND IMPACT INCREASES FOR BOAT **C** THAT ANGLES INTO THE WIND. FOR THIS REASON, A SAILCRAFT ATTAINS MAXIMUM SPEED WHEN DIRECTED AT AN ANGLE UPWIND RATHER THAN CROSSWIND OR DOWNWIND. IT CAN'T SAIL DIRECTLY UPWIND, BUT IT CAN SAIL TO A DESTINATION UPWIND BY ZIG-ZAGGING BACK AND FORTH. THIS IS CALLED *TACKING*.

CONCEPTUAL Physics

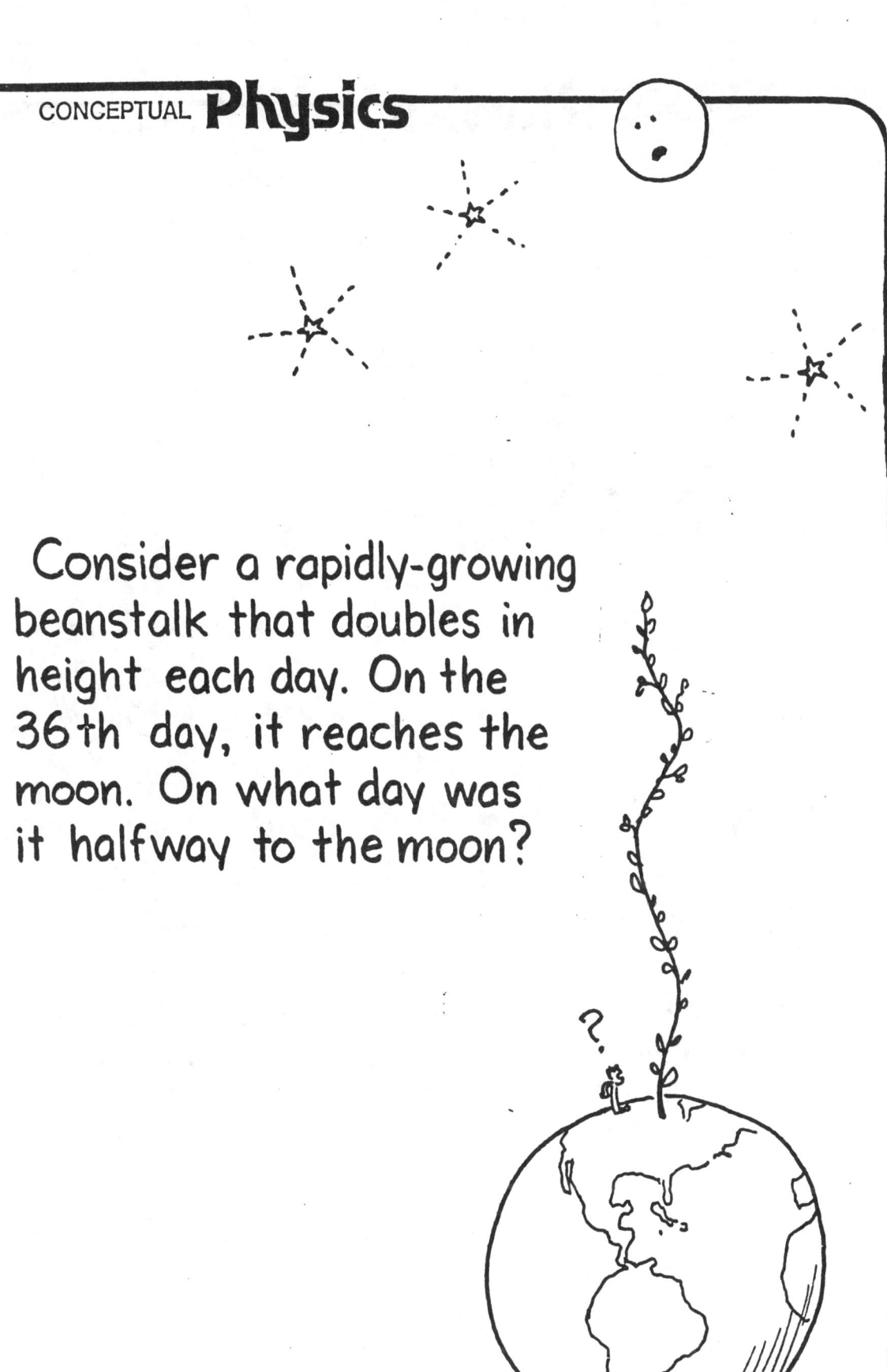

Consider a rapidly-growing beanstalk that doubles in height each day. On the 36th day, it reaches the moon. On what day was it halfway to the moon?

CONCEPTUAL Physics

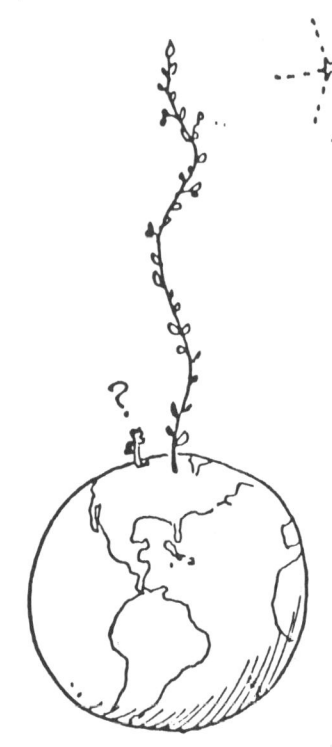

Consider a rapidly-growing beanstalk that doubles in height each day. On the 36th day, it reaches the moon. On what day was it halfway to the moon?

Answer: On the 35th day!

If the beanstalk doubles in height each day, then working backward, its height each previous day must have been half. So on the 35th day the beanstalk must have reached halfway to the moon (and on the 34th day, half again or one quarter as far, and so on).

Keep halving and you'll find that on the first day the beanstalk was slightly more than 1 cm tall!

APPENDIX IV

CONCEPTUAL Physics

Most of us know that when a sheet of paper is folded in half three times, the thickness of the wad of paper is 2 x 2 x 2 or 2^3 or 8 times the thickness of a single sheet. The thickness of a typical sheet of thin paper is about 7×10^{-5} m.

If you could fold a sheet of paper in half 51 times, the height of the resulting "pile of paper" would

a) be about 1 kilometer.
b) reach the moon.
c) reach the sun.

CONCEPTUAL Physics

Most of us know that when a sheet of paper is folded in half three times, the thickness of the wad of paper is 2 x 2 x 2 or 2^3 or 8 times the thickness of a single sheet. The thickness of a typical sheet of thin paper is about 7×10^{-5} m.

If you could fold a sheet of paper in half 51 times, the height of the resulting "pile of paper" would

a) be about 1 kilometer.
b) reach the moon.
c) reach the sun.

The answer is c:
The "pile of paper" would reach from the earth to the sun!

Calculation: For 51 foldings we must multiply the thickness of the paper by $2^{51} = 2.3 \times 10^{15}$.

$(7 \times 10^{-5} \text{ m})(2.3 \times 10^{15}) = 1.6 \times 10^{11}$ m.

Approximation: Every ten doublings multiples the thickness by a factor of $2^{10} = 1024 \sim 10^3$. Fifty doublings will multiply the thickness by

$2^{50} = (2^{10})^5 = (10^3)^5 = 10^{15}$.

Fifty-one doublings multiplies the thickness by a factor of approximately 2×10^{15}. When this is multiplied by the thickness of one sheet the resulting thickness of the pile of paper is 1.4×10^{11} m.
(The distance from the earth to the sun is 1.5×10^{11} m.)

Question: If the original sheet of paper had an area of 600 cm² (roughly the area of a sheet of notebook paper) and if the resulting pile of paper was a square in cross section, what would be the width of the pile of paper?

Answer: The volume of the original sheet is $(6 \times 10^{-2} \text{ m}^2) \times (7 \times 10^{-5} \text{ m}) = 4.2 \times 10^{-6}$ m³. If we assume that the volume is conserved in this redistribution of paper, then the cross-section area of a column reaching the sun is 4.2×10^{-6} m³/1.6×10^{11} m or 2.6×10^{-17} m². The edge of a square of this area is 5.1×10^{-9} m. This is roughly 50 times the diameter of the hydrogen atom.

thanx to Al Bartlett